Dialogue With Our Genes

Ehud Banne • Ben-Zion Shilo

Dialogue With Our Genes

What Our Genes Tell Us and What We Tell Them

Springer

Ehud Banne
Head of Department,
Institute of Human Genetics
Wolfson Medical Center
Holon, Israel

Ben-Zion Shilo
Department of Molecular Genetics
Weizmann Institute of Science
Rehovot, Israel

ISBN 978-981-95-9570-9 ISBN 978-981-95-9571-6 (eBook)
https://doi.org/10.1007/978-981-95-9571-6

This Springer imprint is published by the registered company Springer Nature Singapore Pte Ltd.
The registered company address is: 152 Beach Road, #21-01/04 Gateway East, Singapore 189721, Singapore

Acknowledgments

The creation of *Dialogue with Our Genes* represents an immensely productive and enjoyable journey of exploration and conversation on multiple levels. At its core is the ongoing dialogue that unfolded between Dr. Ehud Banne and Prof. Benny Shilo until Prof. Shilo's death in 2025, shortly before the book was published in its final form. This book is dedicated to Benny.

Benny was fascinated by how exponential growth in genetic knowledge has broadened the scope of genetic counseling. This progress enables precise identification of genetic syndromes down to individual genes and even single nucleotide variations. In light of such rapid advancements, fostering meaningful dialogue about genetics has become increasingly crucial.

Benny was deeply committed to expanding the frontiers of basic genetic knowledge, and he was excited about sharing his passion with the broader public. Readers are encouraged to visit http://bennyshilo.com to view his photography gallery and to view "Benny Shilo – A Life Well Lived" (https://www.youtube.com/watch?v=50cK5l_zFsI), a video that he created shortly before his passing.

Before he left us, Benny expressed deep gratitude to his own family members for sharing the personal story that sparked the book's creation.

Ehud would like to express his deep gratitude to those who have supported and shaped his journey. To his wife, Avital, and his children, Mica and Yiftach, for their patience, love, understanding, and unwavering support. To his parents and brother, for shaping his personality and beliefs. To the professors who guided him, and to the dedicated doctors he had the privilege of working with at Hadassah and Kaplan Medical Centers. And to the talented team he is honored to lead today at the Department of Medical Genetics in Wolfson Medical Center, for their commitment and collaboration that inspire him every day.

Ehud would also like to express his genuine appreciation for the wonderful support from Benny's daughters Anat and Ronie and son Avner, and for their invaluable assistance in completing this book and guiding it through the final stages of its journey.

A vital dimension of the genetic dialogue occurs between genetic counselors and their patients. Counselors must translate complex genetic analyses into clear, actionable information, empowering patients to actively participate in decision-making. These decisions are often influenced by the patient's cultural and religious context, especially when multiple viable options exist. As analytical and diagnostic tools continue to evolve, the patient's role in these conversations will become even more prominent.

This book has two purposes. First is to enhance patients' understanding and awareness, equipping them to make informed choices that could shape future generations. Second, the authors hope to highlight how important it is to support patients by providing clear scientific explanations and helping them feel actively involved in the decision-making process.

Additionally, this book emphasizes the value of the information that patients share with us. This information becomes extremely useful in the identification and treatment of future genetic syndromes that exhibit similar features. In addition, the formation of disease-oriented family groups not only provides moral support but also is an effective avenue for processing information and its use for medical treatment.

The hallmark of this book lies in its integration of diverse languages and disciplines into a cohesive narrative. This achievement owes so much to the professional, insightful, and occasionally humorous editing of Sandra Cash. Sandy's exceptional ability to craft a flowing, accessible dialogue without compromising scientific depth and rigor was invaluable. Her remarkable skills in presenting complex scientific issues in an engaging and comprehensible manner provide the flowing story line of the book.

The editing of this book was generously supported by the Braginsky Center for the Interface Between Science and the Humanities at the Weizmann Institute of Science. By creating a scientific text that is accessible to a broad audience, the book aligns with the center's mission to bridge disciplines and inspire informed decision-making. Moreover, the ethical issues raised within its pages provide a valuable framework for addressing complex decisions that will arise in the future.

The authors would like to thank Prof. Zeev Paroush of the Hebrew University of Jerusalem for his close reading of the book manuscript. They also extend their sincere gratitude to Kumar Athiappan at Springer Nature for his professional and devoted handling of the publication process.

About This Book

Dialogue with Our Genes is a guide to the rapidly changing world of genetic counseling. The book follows two scientists, one a laboratory researcher and the other a clinician, who decided to cross the bridge between them and discuss the impact of science-driven advances on the identification and treatment of genetic disorders.

With the growing availability of vast amounts of information about our genes, more and more families are seeking answers about themselves and their unborn children. This has transformed genetic counseling from a niche specialty to a field that many people will encounter over the course of their lives.

In this book, you will read clinical case studies inspired by patients from Dr. Ehud Banne's practice. Commentary from Professor Benny Shilo will help readers understand the fundamentals of genetics as well as the exciting developments in scientific technologies that have health implications for us and for our offspring.

Dialogue with Our Genes explores the dynamics of medical genetics and the important role played by scientists, genetic counselors, and parents in making informed decisions about each family's future through cutting-edge technology, clinical acumen, and compassion.

Contents

1 An Unexpected Reunion, and the Dialogue Begins ... 1

2 FAQs: The Genetic Counseling Experience ... 5
2.1 How Do I Make an Appointment with a Genetic Counselor? ... 5
2.2 What Does an Ideal Genetic Counseling Facility Look Like? ... 6
2.3 How Do I Prepare for the First Consultation with the Counselor? ... 6
2.4 What Will My First Meeting Look Like? ... 6
2.5 How Does the Counselor Determine Which Tests Should Be Done? ... 7
2.6 How Long Does It Take to Get Genetic Test Results? ... 7
2.7 What Can I Learn from My Genetic Test Results, and How Are Results Communicated? ... 8
2.8 How Can I Ensure My Test Results Remain Private? ... 9
2.9 What Personal Qualities Should I Be Looking for in a Genetic Counselor? ... 9
2.10 What Happens if Test Results Are Inconclusive? ... 10
2.11 What if the Genetic Counselor Does Not Share My Values? ... 11

3 Speculation, Faith, and Professional Advice: The Long Road to Modern Genetic Counseling ... 13
3.1 Evolving Theories of Inheritance ... 14
3.2 A Dark Detour: Genetics, Ideology, and Social Engineering ... 15
3.3 Historical Emergence of Genetic Counseling ... 16
3.4 Tools and Technology ... 16
3.5 Dysmorphology: Genetic Diagnosis "By Sight" ... 17

4 On Smooth Peas and Healthy Babies ... 19
4.1 DNA: An Introduction ... 20

5 Journey to the Center of the Cell ... 25
5.1 How Scientists Read the Landscape ... 27

6 Seeing the Genetic Forest for the (Family) Trees . . . 29

7 What Flies and Worms Teach Us About Human Disease . . . 35

8 Chromosomes: Packing for the Genetic Journey . . . 41
8.1 Packing the Nucleus . . . 43
8.2 Repacking for Cell Division . . . 44

9 Bringing Diagnosis into Focus: The Problem of Missing and Extra Genes . . . 49

10 The Diagnostic Deep Dive: The New Genetic Technologies Providing Answers . . . 55

11 From "Bruiser" to Prospective Dad . . . 61

12 Genetic Tests During Pregnancy . . . 67
12.1 FAQ: Prenatal Genetic Testing . . . 68
12.1.1 Why Perform Prenatal Genetic Analysis? . . . 68
12.1.2 Who Should Be Tested? . . . 68
12.1.3 What Prenatal Genetic Test Can Be Done Early in Pregnancy? . . . 69
12.1.4 What About Prenatal Genetic Analysis Later in Pregnancy (Amniocentesis)? . . . 70
12.1.5 What About Testing the Umbilical Cord? . . . 70
12.1.6 What Are the Risks of Prenatal Genetic Analysis? . . . 71
12.1.7 How Are Genetic Tests Done During IVF, Prior to Embryo Implantation? . . . 71
12.1.8 Who Makes the Decision About testing, the Parents or the Doctor? . . . 73

13 Same Syndrome, Different Symptoms, and the Case of the Hidden Disease . . . 75

14 Out of the Blue: De Novo Mutations . . . 81
14.1 Out of the Blue . . . 85

15 The Chromosomal Origins of Sex-Specific Disease . . . 87
15.1 The Delicate Balance of XX Gene Expression . . . 89
15.2 Syndromes That Skew Male or Female . . . 89
15.3 Caution: Extreme Imbalance . . . 90
15.4 Considerations for Genetic Counseling . . . 91
15.5 Parental Preference . . . 91

16 The Genetics of Gender: Genes, Hormones, and Embryonic Sexual Development . . . 95
16.1 The Hormone Tag Team . . . 96
16.2 Differences in Fetal Sexual Development . . . 97

17 Nature and Nurture Together: The Challenge of Multifactorial Conditions . . . 101
17.1 Gene Plus Gene Plus Gene . . . 104
17.2 Searching for the Multifactorial Origin Story . . . 105
17.3 The Wider Genomic Perspective . . . 106

18 When Genetic Diagnosis Is a Mid-Life Crisis: The Special Case of Late-Onset Disorders . . . 109
18.1 The Impact of a Late-Onset Diagnosis . . . 111
18.2 The Challenge of Treatment. . . . 112
18.3 Living with GNE Myopathy: Resilience in the Face of Decline . . 113
18.4 Late Onset and the Ethics of Pediatric Genetic Testing. . . . 113

19 The Genetics of Inherited Cancer . . . 115
19.1 What Are Tumor Suppressor Genes?. . . . 116

20 Genetic Knowledge, Ethical Dilemmas. . . . 121
20.1 Playing God? . . . 123

21 The Future of Medical Genetics . . . 129
21.1 Genetic Counseling of the Future . . . 129
21.2 Fundamental Genetic Knowledge Set to Increase . . . 131
21.3 A Future (and Necessary) Reduction in Genetic Testing Cost . . . 132
21.4 A Partner for Genetic Progress: Artificial Intelligence (AI) . . . 132
21.5 Waiting in the Wings: New Diagnostic Technologies . . . 133
21.6 Improving Genetic Disease Treatment. . . . 133
21.7 Future Ethical Challenges . . . 134

Glossary . . . 135

Chapter 1
An Unexpected Reunion, and the Dialogue Begins

As I walked into my examination room to begin a first-time clinical consultation, three faces instantly looked up to greet me. The only pair of eyes that didn't seek mine were those of my newest patient—a sweet little girl, perhaps 3 years old, who showed clear signs of delay in her development. As I entered, the child rushed to hide her face in her mother's lap, apparently overwhelmed by this change to what was already an unfamiliar environment.

The scene before me was unusual. As a physician specializing in clinical genetics, I naturally assume that my youngest patients will arrive at my office together with a parent or, ideally, two. But this particular child—blonde, neatly dressed—had an augmented support team of three: the mother, the father, and an older gentleman who looked strangely familiar.

After introducing myself, I learned that the little girl's name was Shiri and that her parents were Keren and Yoav. But just as Keren began to introduce me to her father, a memory shot to the surface.

"Professor Shilo?" I asked.

My hesitant half smile was answered by the older gentleman's wider smile, and a knowing glance from his daughter. She was clearly used to her father being recognized, at least in certain circles.

"Please. Call me Benny," he replied.

Well, I thought, this would take some getting used to. Back when I was a master's student, I attended a lecture on the genetics of embryonic development given by Professor Ben-Zion Shilo—Benny. He was one of the founders of the field and very much admired. At the time, I didn't dare approach him; my own research wasn't going very well, and I was beginning to doubt my potential in the world of science. Now, however, with a medical degree and further training in pediatrics and genetics under my belt, I was working as a clinical geneticist, a field that was in constant flux, based on advances made in basic science labs like Benny's. A one-on-one conversation with a scientist of Benny's stature was a great opportunity.

But as much as I would have liked to take advantage of Professor Shilo's presence to talk shop, the trepidation radiating from both young parents' eyes reminded me that my clinical responsibilities came first. This family had come to talk about Shiri, and that's exactly what we did.

I turned to Keren and asked, "What brings you to the clinic today?"

"Shiri had a normal birth and appeared to be a healthy baby at first," Keren began. "However, at around the age of one, we noticed that she was developmentally delayed

E. Banne, B.-Z. Shilo, *Dialogue With Our Genes*,
https://doi.org/10.1007/978-981-95-9571-6_1

compared to other children her age. She didn't reach milestones like sitting, standing, or walking on time. By the time she turned three, her speech was also delayed, and we were told that she was on the autism spectrum."

While Keren spoke, Yoav pulled a file out of his backpack. It contained all the evaluations done by Shiri's physical, occupational, and speech therapists, as well as a recommendation from a developmental specialist that the couple seek genetic testing.

"Our doctor explained that the genetic testing might not change the diagnosis or the treatment she's receiving," Yoav said. "But we want to understand what's behind Shiri's developmental delay and get a clearer picture of what we can expect in the future."

Just a glance at the thick file chronicling Shiri's therapeutic history revealed that Keren and Yoav had noticed their daughter's delays early on and had taken quick action to ensure appropriate intervention. I also thought it safe to assume that as the daughter and son-in-law of a geneticist, the couple might be ahead of the game, understanding better than most of my patients what clinical genetics can and cannot do. Luckily, however, my training kicked in, and I set aside my assumptions so that I could be completely open to meeting this family where they were and gather the information needed to guide them through their medical crisis. Of course, that didn't mean that I didn't look forward to striking up a personal conversation with Professor Shilo—excuse me, Benny—sometime in the future.

As in any first meeting with new patients, I concentrated on getting to know the family, investigating if there were any known genetic syndromes among their relatives, and explaining the nuts and bolts of the genetic testing process. Handing them the forms needed to order blood samples from both Shiri and her parents and begin the genetic sequencing process, I bade the family farewell for now and promised to keep them updated. Then something unexpected happened.

On his way out, Benny paused at the door and turned to me. "Dr. Banne," he began. My smile couldn't be contained. "Please, call me Ehud," I replied.

"You know that I have been a professor at the Weizmann Institute for decades, and I have been studying genetics for my entire adult life," Benny said, "It is ironic that this treatment for Shiri—my own granddaughter—is my first exposure to the process of doctors identifying genetic syndromes in patients. I completely trust your expertise, but I want to offer my help if there is anything you think I can assist with. If you would be open to working together, we might find that we have a lot to learn from each other."

With that, Professor Shilo—or, rather, Benny—handed me his business card, smiled, and left.

What an amazing turn of events! I took a moment to reflect on this opportunity to expand, with Benny's help, my knowledge and understanding in fundamental genetics, which is crucial to my role as a physician and geneticist. Shaking my head in wonderment, I sat down at my desk, scanned my schedule, and reminded myself that Shiri was only the first of many patients I would be seeing that day.

Benny and Ehud had crossed paths many years ago, though Benny hadn't been aware of it at the time. Now, the two had reunited at a different, and very dramatic, period in the history of genetic science.

The 2001 completion of the Human Genome Project, which sought to identify and decipher the genetic code contained in human DNA (or deoxyribonucleic acid), marked a turning point in the ability to use genetic information to make informed medical decisions. In the decades since, advances in scientific technology have made it possible to extract more detailed genetic information than ever before.

The impact of these advances can be seen in the cost and speed by which DNA can be analyzed. The Human Genome Project took 15 years to complete, involved teams from around the world, and cost $2.7 billion. Today, sequencing—that is, determining the exact order in which the fundamental components of a DNA molecule appear—takes just days and can be completed at a tiny fraction of that cost. To put the savings in perspective, this is like a commercial airliner that cost hundreds of millions to build going on sale for $100.

The increased accessibility to DNA data via sequencing and other advanced technologies has opened a new way of communicating with our genes: interrogating them for useful information that can have an impact on our health and well-being. But this has an added interpersonal component when medical experts communicate gene-related data to families. This is what Ehud, the pediatrician and clinical geneticist, focused on during his first extended conversation with Benny, the laboratory scientist.

As it happened, Benny called me first. The results of Shiri's sequencing had not yet come in, but as a scientific investigator who had spent his career chasing down the answers to interesting questions, Benny was curious to learn about my practice, which seemed so related to, yet so different from, his own research. What is the key, he asked, to successful genetic counseling, and how do I approach my patients?

"A genetic counselor's job is all about communication," I began, slipping easily into the explanation I had often given to medical students, while simultaneously marveling at the turn of events that had put me—of all people!—into this advisory role. "Our goal is to help patients and their families understand and adapt to the medical, psychological, familial, and reproductive implications of genetic inheritance. The best-case scenario would be to find the cure for a condition based on patients' genetic information. But this is not always possible; there are many different kinds of families, and most of the genetic challenges they face have more than one potential solution."

"I certainly feel challenged," Benny replied. "Obviously, the situation is hard on the whole family, but for me, it's personally frustrating. I know how genes work in cells and in model systems, but I don't have the tools to diagnose the condition that affects my own granddaughter."

"Look at it this way," I said. "Geneticists like me would never be able to do their jobs without lab-based technologies, which are improving all the time. There's a lot we still don't understand about hereditary disease, but thanks to science, our genes are becoming more of an open book."

"But here's the thing," said Benny. "As scientists, we look for definitive answers. But your clinical work seems more interpretive, giving me the impression that genetic counseling is an art as well as a science."

"I tend to agree," I said. "Identifying a gene at the root of some condition or syndrome is rarely the end of the story. Genetic diagnosis is just the beginning of a process that carries emotional weight. That's one of the most fascinating things about clinical work: as a geneticist in clinical practice, my work is based on science. But once we've clarified the facts, there may be many options for moving forward. We don't pass judgment; what families choose to do with the genetic information we give them necessarily depends on individual preferences and cultural values."

"This is exactly what I find so fascinating, because I seem to be wearing two different hats," Benny said. "As a grandfather, all I want is for Shiri's sequencing to come back and reveal something that will help us give her the best possible life. At the same time, I'm a scientist. I've been investigating the nuts and bolts of genetics for as long as I remember. I'm curious to know how the science affects what you do and how your work affects all your patients."

"Rest assured, curiosity is a two-way street," I said. "Look, here's a proposal: as we go through this process, let's stay in touch. I would welcome the opportunity to learn more about your research and see what's happening at genetic science's cutting edge."

"I think this is a dialogue with potential," he agreed.

"Maybe you'll even get a publication out of it," I said with a smile, indicating I had made a joke, one that I hoped Benny would sense over the phone.

"Maybe we both will," he replied. And though I couldn't see it, I knew Benny was smiling too.

Chapter 2
FAQs: The Genetic Counseling Experience

Before we begin this journey, we present some frequently asked questions about genetic counseling that may help orient readers to how these dynamics play out in clinical settings from the patient's perspective and provide some practical, useful information.

Imagine that a close relative has just been diagnosed with cancer. Or perhaps you're pregnant and your doctor has ordered genetic tests but you don't really understand why. Perhaps you're considering starting a family but you're worried about conditions that may run in your family. These are all good reasons you might want to meet with a specialist in medical genetics, such as a clinical geneticist (a physician) or a genetic counselor, depending on the healthcare system and local practice. But how can you arrange such a meeting, and what should you expect? These are the questions that this chapter seeks to answer.

2.1 How Do I Make an Appointment with a Genetic Counselor?

Most genetic counseling sessions begin with a referral. Speak to your family doctor or a specialist and explain your concerns. Ask for a recommendation for the best place to get genetic counseling in your area (it could be in a hospital or a private clinic).

E. Banne, B.-Z. Shilo, *Dialogue With Our Genes*,
https://doi.org/10.1007/978-981-95-9571-6_2

2.2 What Does an Ideal Genetic Counseling Facility Look Like?

A good counseling session begins with the room itself. It should be quiet and private, a place where you can speak without hesitation and know your words won't be overheard. The space is arranged to make conversation easy: comfortable chairs, little clutter, phones switched off so attention isn't divided. For patients who need it, there is access for people with disabilities, materials in the language they prefer, and a translator who can be present throughout. Counselors often use simple tools—a screen to show results, a handout to take home, or even a whiteboard to sketch out complicated ideas. These small details matter. They tell the patient that the focus is on them, that they are being listened to, and that the discussion about their genetic results and choices is meant to be as clear and as supportive as possible.

2.3 How Do I Prepare for the First Consultation with the Counselor?

Work with your general practitioner or family physician to collect a complete set of healthcare records, along with the results of any tests or developmental workups performed in the past. As genetic counselors will want to know about your family, try to collect medical information about your family members as well. Keep in mind that the decision to seek genetic counseling can be stressful; it can be helpful to write down your questions and bring a notebook for writing down the answers you receive.

2.4 What Will My First Meeting Look Like?

The genetic counselor will review any healthcare records you bring and will examine the individual being investigated for the presence of a genetic syndrome. Although every case is different, most intake examinations involve three general strategies: an interview to establish the family's concerns and the patient's relevant health history; a physical examination (typically performed by a clinical geneticist or appropriately trained clinician) to look for signs of a genetic condition; and an investigation of the family's history of relevant health conditions. This investigation be used to create a family tree showing how the suspected syndrome may have been passed down through the generations, while highlighting other medical information which may be of importance.

Based on the examination, the counselor may order genetic tests to confirm whether a suspected syndrome is present and to identify its cause. These may include targeted assays, which look for a specific known mutation; chromosomal microarrays, which scan all the chromosomes to detect missing or extra pieces of DNA; and next-generation sequencing (NGS), which reads large sections of DNA or RNA to uncover more subtle changes in the genetic code.

2.5 How Does the Counselor Determine Which Tests Should Be Done?

The information gathered during the first meeting—your health history, your family's history, and the findings from a physical exam—guides the counselor toward a working idea of what condition may be present. The test is then chosen to answer that specific question. Sometimes this means using a targeted assay, which looks only for a single known mutation; sometimes it involves a chromosomal microarray, which scans all the chromosomes for missing or extra pieces of DNA; and sometimes it requires next-generation sequencing (NGS), which rapidly reads millions of small pieces of DNA or RNA and pieces them together to reveal more subtle changes in the genetic code.

2.6 How Long Does It Take to Get Genetic Test Results?

Unlike scientific research labs, where investigators may spend years trying to solve a particular genetic mystery, clinical genetics labs face time constraints. Waiting years for results is not an option, and patients expect clear, definitive answers to a critical question: does the patient have a particular genetic syndrome or not?

The time lapse between the initial referral to genetic counseling and a family's first appointment at the clinic, and between ordering genetic tests and getting test results, can be challenging for both the patient and counselor. The speed at which this process plays out also varies between countries and between urban centers, which often benefit from the presence of a strong medical infrastructure, and the periphery.

Israel, where Benny and Ehud live, is home to one of the world's most advanced medical systems. At the time of this writing, however, it can take a year for a patient referred to genetic counseling to get an appointment at the clinic. Timelines vary widely by healthcare system.

Once it is determined which genetic tests are needed, it can take up to 2 months for a patient's health maintenance organization (in Israel, there is universal medical insurance by law, but citizens choose among competing HMOs) to approve coverage of the testing cost.

Then it can take another 3 months to get an appointment for the test. The time it takes for test results to be fully analyzed and returned to the genetic counselor for communication to the patient varies from test to test.

Again, the above timeline is accurate for the Israeli medical system at the time that this book was written. Depending on location, families who are referred to genetic counseling may have a faster or slower experience.

2.7 What Can I Learn from My Genetic Test Results, and How Are Results Communicated?

Once the tests are completed, the lab interprets the findings and reports the results directly to the genetic counselor, who will then communicate them to the patient or family. This is to prevent patients or their families—who may lack the professional training to understand test results without the counselor's guidance—from drawing erroneous conclusions.

Having the genetic counselor present test findings rather than the lab is important for another reason. Patients expect clear, definitive answers—positive or negative. This necessity for decisiveness forces labs to make firm conclusions, even in the face of uncertainty. Labs will sometimes communicate the uncertainty in their reports. Aware of this uncertainty, the counselor draws on their clinical training to navigate questions such as the following: Do the test results reflect the patient's condition? Can the genetic findings affect the patient's future health or the health of future offspring? How do these variables impact the genetic counseling process? What does the family really need to know?

Patients often fear the worst, and counselors may prefer not to handle the emotional burden of delivering bad news alone. In some cases, counselors request that another qualified professional, such as a nurse, social worker, or fellow genetic counselor, be present to provide support.

Most counselors prefer to avoid disclosing test results over the phone, opting instead to invite the patient for an in-person session where the findings can be thoroughly explained. At this stage, patients may be introduced to an unexpected diagnosis or line of future investigation. They often need time to process this information, and the meeting provides a framework for questions, discussion, and emotional support.

It is also important to note that counselors and labs sometimes face dilemmas that make it best to withhold uncertain findings and, instead, focus only on those findings associated with actionable, clear results. For example, if testing reveals that a patient may have heightened risk for a future disease condition that can be verified with a simple medical check, such as a blood test or echocardiogram, it will likely be reported to the patient; otherwise, it may not be.

2.8 How Can I Ensure My Test Results Remain Private?

Genetic testing has clear clinical benefits to the patient. However, it also introduces new risks to patient privacy, and patients are potentially vulnerable to the misuse of their genetic information. Misuse of genetic information can adversely affect a person's access to healthcare, employment, and other rights.

To address this, in 2008, the US Congress passed the Genetic Information Nondiscrimination Act to restrict the access of issuers of health insurance and employers to individuals' genetic information and to prohibit genetic discrimination. Concerns remain as to the use of genomic information to discriminate against applicants for long-term care, life, and disability insurance.

2.9 What Personal Qualities Should I Be Looking for in a Genetic Counselor?

Depending on where you live, it may not be possible to shop around for a genetic counselor. But whoever you ultimately work with, your counselor will not be just a medical professional; they will also be your guide as you embark on potentially emotionally challenging journey. Ideally, your counselor should have a high level of professional training, an empathetic and nonjudgmental outlook, and excellent communication skills.

Explaining complex genetic information requires careful thought and clarity. As with any intricate subject, breaking down the information into manageable, bite-size pieces helps patients better understand their condition. Once the basics are explained, counselors can introduce the more nuanced aspects of the diagnosis.

Genetic counseling involves emotionally charged questions related to diagnosis, prognosis, testing options, and practical considerations. Patients are rightly concerned about the implications of their condition, the chances it could be inherited by their offspring, and the preventive measures available. Often, patients struggle to articulate their concerns during the session and may later remember questions they forgot to ask. These queries are deeply intertwined with emotional responses, which can include a mixture of fear, guilt, and hope.

To support informed decision-making, counselors must be able to tailor their communication to each family's unique needs. Clear, empathetic communication and active listening help reduce anxiety and build trust. Counselors should connect patients to resources, such as support groups and financial assistance, that can empower them to navigate challenges. Allowing patients and families the time they need to process information—sometimes leading to a follow-up visit—can further support their emotional needs.

In situations where emotional or cognitive factors impair decision-making, counselors may need to intervene by offering additional resources or suggesting options that align with the patient's best interests. Striking a balance between respecting autonomy and offering guidance requires sensitivity, effective communication, and a deep understanding of the patient's values and concerns.

In the rapidly evolving field of genetics, where new knowledge is constantly emerging, a good genetic counselor stays up to date with the latest research and clinical guidelines. They also anticipate what patients might expect from the session, whether it be a diagnosis, a sense of closure, or reassurance for future pregnancies. Understanding these emotional and informational milestones allows counselors to approach the individual session more effectively.

2.10 What Happens if Test Results Are Inconclusive?

Genetic counselors frequently need to grapple with incomplete or unclear lab results, such as genetic variants of uncertain significance. Drawing on both their professional training and their own interpersonal sensitivity, counselors must decide how much information to disclose, balancing the need for transparency with the potential to cause unnecessary anxiety. An ability to provide clear explanations without overwhelming patients with technical details is crucial for effective counseling.

When results are inconclusive, the counselor faces the ethical responsibility of balancing a patient's right to know with their capacity to bear uncertainty—a delicate task that calls for sensitivity, honesty, and moral clarity. Counselors must assess whether sharing partial results might lead to premature or inappropriate medical actions, such as unnecessary surveillance or interventions. Additionally, there is an ethical responsibility to revisit and reevaluate uncertain results as new evidence emerges. This requires clear communication with patients about the potential for updates in their genetic information.

The psychological impact of uncertain results can trigger anxiety, fear, and frustration in patients. They may feel helpless when confronted with results that lack immediate answers, particularly when these findings have potential implications for their or their family members' health. Some individuals may experience information paralysis, where the ambiguity of results hinders decision-making. Others may seek premature or unnecessary medical interventions to gain control. In such cases, it is the counselor's responsibility to help clients recognize their own emotional response—if necessary, directing them toward resources for psychological counseling—so that patients and their families can make the decisions that are right for them.

2.11 What if the Genetic Counselor Does Not Share My Values?

Cultural, psychological, and familial dynamics play a significant role in how genetic test results are interpreted. Counselors must navigate these challenges sensitively to maximize the patients' autonomy and informed decision-making, while avoiding directive action rooted in personal values to which the patients may not ascribe.

In some cultures, for example, familial duty, collective decision-making, and the stigma of genetic conditions influence how patients and their families respond to uncertain findings. In traditional East Asian societies such as China, Japan, or Korea, collectivism and filial piety are emphasized. In these cultures, the well-being of the family often takes precedence over individual autonomy. Uncertain genetic results might be viewed through the lens of their potential impact on future generations rather than on the individual.

Some families may prioritize understanding the implications of genetic findings for children or grandchildren, even if immediate concerns for the individual are minimal. This cultural perspective can lead to pressure to take premature action, such as pursuing further testing or treatments, to safeguard the family's health, despite the individual's preference for waiting for more definitive information. Concerns about genetic conditions being seen as hereditary "flaws" could result in attempts to control how this information is shared within the family or community, reflecting a broader cultural emphasis on maintaining family honor.

In extreme Orthodox Jewish communities, where marriage partners are chosen by external agents, genetic counseling carries unique cultural significance. Genetic information can significantly influence the matchmaking process, as community *shadchanim* (marriage brokers) gather detailed information about prospective brides and grooms, including health status. When uncertain results are found, families may face significant cultural and psychological challenges. For instance, carrier status for a condition may raise concerns about how it will affect an individual's marriage prospects or even the prospects of distant relatives, even if the condition's immediate health implications are minimal. Families might feel torn between disclosing information that could help safeguard the health of future generations and withholding it to avoid jeopardizing the individual's "value" on the marriage market. The pressure to preserve familial reputation in these communities can complicate decisions about disclosure and confidentiality.

Furthermore, in other cultural contexts, patients may view uncertain genetic findings as a personal burden or a potential source of shame, making them reluctant to share information with family members. Language barriers and varying levels of health literacy add further layers of complexity, requiring tailored communication strategies to ensure comprehension.

The counselor must navigate these emotional responses with empathy and sensitivity, providing reassurance and fostering a collaborative approach to decision-making. Counselors should explain the benefits, risks, and uncertainties of different options while remaining nondirective and giving patients the support they need to make their own independent decisions and maintain a sense of control.

Genetic counselors often face challenges balancing a nondirective stance with meeting patient needs, particularly when patients explicitly seek advice. This tension is heightened by the complexity of risk communication, where differing perceptions of risk can subtly influence decision-making. The evolving nature of genetic testing, combined with cultural factors, requires the counselor to be flexible about the communication strategies they use to fulfill patients' needs while respecting their values. The counselor must avoid personal judgment and refrain from directing the patients. Indeed, nondirectiveness is one of the basic rules of genetic counseling.

As Benny learned, working with patients in a genetic clinic is nothing like leading a team of genetic scientists in a lab.

Benny: In my research lab, even though we all work together, everyone accepts that I have a broader role in guiding the decision-making, based on my years of experience. Do you ever find it difficult to put your preferences aside and let patients decide?

Ehud: As a director of a genetic institute, I definitely guide my team. But in front of patients, that's different. Counselors often face challenges balancing a nondirective stance with meeting patient needs, and that's why there has been academic research, by Tina-Marie Wessels and others, about helping genetic counselors do their job faithfully while dealing with the stress. Supervision and mentorship can provide perspective and help counselors manage the stress of their patients' emotional response with professionalism and greater confidence.

Benny: When test results are inconclusive, do you find it difficult to withhold information about what the tests might mean?

Ehud: Throughout my career, I have found that no patient leaves a session with everything fully clear. Even physicians and PhDs in genetics often have lingering questions about their condition and need additional time to process what they've learned. Over time, most patients will come to understand what is most important to them, and they will believe they know all they need to know. However, as a scientist, I know that new knowledge is always emerging. This is why annual follow-up appointments are crucial, typically a year after the diagnosis. There's always the chance that a variant of unknown significance—what we doctors call a *VUS*—that appeared in the original test results will later be reclassified as either harmless or disease-causing, altering what it means for the patient's care and future.

Benny: Sounds like genetic counseling is not what you'd call an exact science.

Ehud: The role of genetic counselor goes beyond simply explaining test results. We act as advocates for patients going through a difficult process, helping them navigate the gap between science and their own lives. In the best of cases, we succeed in helping our patients make informed decisions through trust, empathy, and clarity. It's all about knowing that even in the most difficult situations, there is guidance, support, and hope.

Chapter 3
Speculation, Faith, and Professional Advice: The Long Road to Modern Genetic Counseling

Human fascination with the mechanics of heredity—how traits are passed down from one generation to the next—goes back to ancient times. We have good empirical evidence that humans were breeding plants and animals for desired characteristics as early as 15,000 BCE, based on traces of agricultural products buried in sealed tombs. This is also supported by the genetic profile of man's best friend. Thanks to modern genetic techniques, we know that the domesticated dog descended from wolves selected by their ancient owners for behavioral traits like herding, guarding, and hunting.

In the fifth century BCE, Greek physician Hippocrates proposed a theory of heredity that involved the transmission of physical material, in the form of semen, from the father to the child. This theory, which was likely developed by Hippocrates himself, became known as the *bricks and mortar* model.

Aristotle's ideas on heredity, proposed a generation later, offered a different perspective. He observed that even physically handicapped men could father children with normal physical traits and proposed that heredity was instead based on the transmission of information. This idea, known as the Aristotelian blueprint model, closely resembles our modern understanding of genetics. First proposed by Aristotle in the fourth century BCE, the notion was ignored for over 2000 years, only to reappear in the nineteenth century, revived by the discoveries of Gregor Mendel and the dawn of modern genetics.

Because this book has been written by a scientist and a physician-geneticist who both live in Israel, it is perhaps most fitting to begin our discussion of heredity with the Hebrew Bible. The Bible provides clear evidence that our ancient ancestors were aware that hereditary traits are passed down through the generations, even if our biblical forebears did not display a deep understanding of its underlying mechanics.

In the Book of Genesis, the patriarch Jacob—a shepherd, like Abraham and Isaac before him—manipulated hereditary principles for his own advantage. After working for his uncle and father-in-law Laban for 14 years, Jacob was ready to take his

E. Banne, B.-Z. Shilo, *Dialogue With Our Genes*,
https://doi.org/10.1007/978-981-95-9571-6_3

wives and children home to Canaan. In a bid to retain Jacob's shepherding services, Laban proposes to let Jacob name his wages. Jacob responds that he is willing to stay on and tend Laban's flock for what seemed like a paltry fee, one that could not possibly be cheated on:

"Let me pass through all your flock today, removing from it every speckled and spotted sheep, and every black lamb, and the spotted and speckled among the goats, and they shall be my wages." (Genesis 30:32)

In agreeing to these terms, Laban may have believed he made a wise business decision. However, he didn't know that Jacob was an astute observer of sheep-breeding practices. Jacob intuited that his future wealth depended on getting spotted and speckled animals to outbreed the single-color animals.

In the biblical narrative, it is suggested that the proportion of spotted versus pure-colored animals is influenced by divine intervention rather than genetics: God appears to Jacob in a dream and shows him that he will prosper if he encourages the breeding of spotted animals. Whether or not this dream provided Jacob with the knowledge to identify the potential for spottedness in sheep and goats that were initially pure white or black, his breeding efforts resulted in a significant increase in the number of spotted animals in Laban's flock. Under Jacob's careful management, the number of spotted sheep and goats grew much faster than anticipated, making Jacob, who claimed ownership of these animals, a wealthy man. There can be no doubt Jacob was aware, at least partially, of some laws of inheritance.

Despite the biblical patriarch's early insights, it took millennia for animal breeding to become a science. Many of the most important developments along this road will be described in subsequent chapters. But to fully grasp their revolutionary impact, it will be helpful to trace briefly the chronology of humankind's understanding of heredity.

To begin this story, let's jump from the Bible to the dawn of the modern age.

3.1 Evolving Theories of Inheritance

Prior to the mid-nineteenth century, the most widely accepted concept of heredity was the *blending theory*, the idea that offspring's characteristics are equally influenced by both of their parents. A studious Austrian monk named Gregor Johann Mendel (1822–1884) challenged this conception with a whole new approach.

Mendel's groundbreaking study of thousands of pea plants explained why certain traits that seemingly disappear in progeny can reemerge in subsequent generations, following a clear mathematical formula. Despite the significance of this discovery, Mendel's insights remained largely unknown until the early twentieth century, when the Royal Horticultural Society of England published an English translation of his 1866 paper. The paper caught the attention of British biologist William Bateson (1861–1926), who was deeply impressed by its findings. Bateson became an influential champion of Mendelian principles and, in 1905, coined the term *genetics* based on the Greek word *genno*, meaning "giving birth." This term soon became the accepted title for the entire field of study.

As the twentieth century unfolded, advances in genetics brought the mechanics of heredity into clearer focus. Early breakthroughs came from the study of inheritance in bacteria and the viruses that infect them. Work by scientists including Max Delbrück, Alfred D. Hershey, and Salvador E. Luria showed that inheritance is mediated by DNA. This led to the discovery of DNA's double helix structure by James Watson, Francis Crick, and Rosalind Franklin. Other scientists—most notably, Robert W. Holley, Har Gobind Khorana, and Marshall W. Nirenberg—uncovered the process by which the information stored in DNA is translated into proteins, the molecules that drive biological function in all living organisms.

3.2 A Dark Detour: Genetics, Ideology, and Social Engineering

Discoveries about the science of heredity have been the source of both significant excitement and significant ethical challenges. To illustrate this, we need to look no further back than the early twentieth century, a dark period for genetic science in which emerging knowledge was weaponized to support a theory of racial purity known as *eugenics*.

The theory of eugenics stated that to purify the population, foreign traits—broadly speaking, traits foreign either to the majority culture or to an authority in power—should be targeted for elimination. This was to be accomplished either through the sterilization procedures that would prevent pregnancy and thus the birth of so-called undesirable children, or through the active removal (i.e., murder) of individuals believed to possess undesirable traits.

The twisted philosophy of eugenics was rooted in the belief that certain populations and traits were better than others and that efforts should be made to preserve "superior" traits and eliminate the "inferior" ones. Laws were passed in several countries, including the United States, that advocated for the forced sterilization of certain groups and restricted immigration based on perceived genetic superiority. These misguided policies laid the foundation for the horrific atrocities committed during World War II by the Nazi regime, where the extermination of "impure" populations, such as Jews, Roma, and others, was a top priority.

The devastating consequences of eugenics have since discredited its principles and relegated it to the fringes of society. However, history serves as a reminder that it is imperative to remain vigilant and not allow genetic science to be co-opted to support any malign ideology.

3.3 Historical Emergence of Genetic Counseling

Genetic discoveries led to positive developments as well. Throughout the twentieth century and beyond, increased understanding and awareness of genetics has had a profound effect on society, influencing decisions about childbearing and childrearing, as well as the medical establishment's response to illness. Over time, this evolution gave rise to an entirely new profession known as *genetic counseling*.

In the 1960s, the field of genetics was still in its early stages, and there was no established profession dedicated solely to genetic counseling. Instead, the diagnostic and clinical care of individuals with genetic syndromes were primarily handled by pediatricians and gynecologists. However, as the understanding of the heritability of certain syndromes grew and the number of referrals for genetic assessments increased, it became clear that a new medical subspecialty was needed. This led to the creation of a specialized field focused on genetics and the establishment of physician-geneticists with interdisciplinary expertise.

Answering the growing need to integrate genetic science into clinical practice, pioneering medical professionals moved the field forward. These include Dr. Victor McKusick, a physician who became known as the "Father of Medical Genetics" after he established a training program for clinical genetics at Johns Hopkins University in 1957. His program, and others like it, empowered doctors—primarily pediatricians and gynecologists—to incorporate genetic testing into their work alongside more traditional diagnostic and therapeutic strategies.

As the demand for genetic expertise grew, a new class of professionals emerged. These were genetic counselors who were not physicians and who received their training in certification programs rather than in medical school. The first such program opened in 1969 at Sarah Lawrence College in Bronxville, New York, under the direction of Dr. Melissa Richter. The program offered a then-unique curriculum that provided students with the medical education they needed to do their job effectively while bridging the gap between scientific genetics and modern healthcare.

Today, the discipline of genetic counseling is well established. It sits at the crossroads between science, clinical medicine, family planning, and ethics.

3.4 Tools and Technology

In the early days of genetic counseling, medical technology was limited, and most of the tests conducted were borrowed from the conventional clinical practice of the time. Most of these tests, such as X-rays, CT scans, blood counts, and thyroid function tests, were not particularly useful for diagnosing genetic disorders. *Karyotyping*, a technique for photographing chromosomes, was one of the few genetic tests available.

Subsequent advancements in technology led to the development of more specific metabolic tests, including the measurement of plasma amino acids, urinary organic acids, and lactate and ammonia levels. While these tests have proven valuable for certain diagnoses, they still have limitations and cannot be used to identify most genetic disorders.

In parallel with these developments, technologies developed for scientific research provided new strategies for identifying genetic conditions and syndromes.

Molecular biologists developed mapping systems that could identify regions of the genome (the complete set of genes or genetic material present in a cell or organism) suspected of involvement in particular syndromes. Taking a more focused approach, a new technology created in the mid-1980s called *polymerase chain reaction*, or *PCR*, made it possible to identify changes within a specific suspected gene. Narrowing the frame further, *Sanger DNA sequencing*, a procedure invented in 1977 and still used for sequencing individual DNA fragments, can define genetic abnormalities at a resolution of a single DNA base, the nucleotides that serve as genetic material's fundamental building blocks. Technologies such as these made it possible for scientists to delve into the information revealed by the Human Genome Project, which in 2001 provided the first complete sequencing of humankind's genetic material.

In the first decade of the twenty-first century, *exome sequencing*—a technique that analyzes only those regions of the DNA molecule that code for proteins—began to generate mountains of data that scientists and medical researchers have only begun to explore. Outside of exomes, there is great variability between individuals' genetic material. *Whole genome sequencing* sheds light on these regions. Scientists believe that mutations in genomes may explain faulty protein function and much more.

3.5 Dysmorphology: Genetic Diagnosis "By Sight"

Alongside these technological advances, a growing community of professionally trained counselors facilitated the emergence of a knowledge-based intuitive approach that remains a crucial component of clinical genetic practice. This approach is called *dysmorphology*.

Dysmorphology involves the visual examination of physical characteristics, such as facial features, eye angle, nose width, and finger shape, that may indicate the presence of a genetic syndrome. Trained to recognize these subtle visual details, experienced dysmorphologists can often diagnose rare genetic disorders immediately.

To understand the specialized skills that an expert dysmorphologist brings to the table, let's stop by the medical genetic clinic where Dr. Banne sees patients and describes the process from his perspective.

> My whole staff knew that when Dr. Eva Gross came to visit, it was time to drop everything, come running, and be prepared to watch, listen, and learn. Dr Gross had already retired when I first met her, but her vast knowledge made her a uniquely treasured figure within Jerusalem's clinical genetics community. Like the rest of the team, I felt lucky when she

> stopped by to consult on tricky cases. Specifically, she came to share her insight as a veteran dysmorphologist, someone capable of identifying suspected genetic syndromes based on visual cues.
>
> On one of these visits, we consulted with Dr. Gross about a child with a severe and uncommon dysmorphology that included a flattened back of the head, wide fontanels (the soft gaps between the bones of a baby's skull), prominent forehead, and widely spaced eyes, whose genetic diagnosis had yet to be clearly established. The case had stumped everyone on our team. Dr. Gross promised to come back in a week with a diagnosis, and she did.
>
> Relying on her own remarkable familiarity with the medical literature, Dr. Gross returned with a diagnosis of Frank-ter Haar syndrome, a condition so rare that our patient was only the 12th child ever diagnosed with this genetic disease. This diagnosis based on dysmorphology was later confirmed by the genetic tests.
>
> The expertise and intuition Dr. Gross brought to bear in constructing this diagnosis made a strong impression on me that day. It made me think—not for the first time—how lucky I am to live in an era where science makes it possible to solve genetic mysteries and offer guidance to patients and their families.

In the following chapters, we'll return to Frank-ter Haar syndrome, which is an excellent example of the multifaceted detective work needed for a clinical team to identify a genetic defect and diagnose a genetic disease. With the help of analytic tools created in scientific research labs, this sleuthing is carried out in stages, with each new stage achieving a higher-resolution understanding of a patient's unique genetic profile.

Chapter 4
On Smooth Peas and Healthy Babies

In the mid-nineteenth century, a monk named Gregor Johann Mendel carried out a set of experiments in the garden of the abbey in Brno, Moravia (present day Czech Republic), and, by doing so, planted the seeds of modern genetic science. At the time, it was known that crossbreeding of plants, and of animals, could give rise to new traits, but the nature of these traits, and the way in which traits were transferred from one generation to the next, was completely mysterious.

Using the humble pea plant, Mendel meticulously observed visible traits such as plant height, seed shape, and color. He began with true-breeding pea strains, in which offspring always inherited the same traits as their parent plants. But when he crossed two different strains, such as those with round versus wrinkled seeds, the next generation of plants revealed a surprising pattern.

To Mendel's surprise, all the first-generation progeny produced only round seeds, seemingly erasing the trait for wrinkled seeds. However, when these first-generation plants were allowed to self-fertilize, one-quarter of the next generation displayed wrinkled seeds, while the remaining three-quarters had round seeds.

In an era when the agents responsible for intergenerational trait transfer—what we now call *genes*—had yet to be discovered, how could such unexpected results be explained? This question spurred Mendel on to develop the laws of inheritance that laid the groundwork for modern genetics.

Gregor Mendel's groundbreaking experiments with pea plants led him to a radical conclusion about the nature of inheritance. Rather than seeking out the physical agents that caused certain traits to be passed down, Mendel considered them as abstract entities. He observed that one property, such as round seeds, was "dominant" over another, like wrinkled seeds. This meant that in the first generation of offspring, only the dominant round-seeded property was expressed, while the recessive wrinkled trait was seemingly hidden. However, in the next generation, when offspring inherited two copies of each trait from both parents, the recessive wrinkles reappeared in one-quarter of the offspring.

E. Banne, B.-Z. Shilo, *Dialogue With Our Genes*,
https://doi.org/10.1007/978-981-95-9571-6_4

Mendel's assumption was that each parent contributed one agent for each property, which combined to determine the traits of their offspring. This notion, though abstract, became the foundation of modern genetics.

To illustrate, one could think of the agents of inheritance like a recipe, with each parent contributing ingredients that determine the outcome of their offspring's traits. Mendel's experiments not only revolutionized our understanding of inheritance but also paved the way for modern genetics research.

This book deals with the genetic information we carry and the ways we transfer it to our children. It is appropriate now to dwell on the biological mechanisms that mediate this intricate process. Let's reexamine Mendel's theory with the view of our current knowledge, a century and a half later.

4.1 DNA: An Introduction

In the heart of every living cell lies the molecule that serves as the blueprint of life: DNA. It is through the genetic information contained within this elegant and intricate structure that all the traits and characteristics that define an organism are transmitted across generations. The genetic information carried by DNA serves as a blueprint for making proteins, the biological molecules that carry out all cellular functions. We will discuss the mechanics of this in more detail later; for now, an introduction to the most important concepts will suffice.

The conveyors of genetic inheritance, called *genes*, are discrete segments of DNA that carry the instructions for building the myriad proteins that govern the function and behavior of our cells. Humans have approximately 20,000 genes, each existing in two copies, one inherited from our mother and one from our father during fertilization of the egg by the sperm. It follows that although we carry genetic information of our parents, we carry only some of that information, and all our features are dictated by the unique blend of parental traits we received.

The process of genetic inheritance is a complex and intricate dance that occurs at fertilization, as the sperm and egg unite to form a new life. The cells of the sperm and egg each contain only half of the genetic information of each parent. The actual composition of that information is random and unique for each sperm or egg.

To understand how this process works, we need to delve into the intricate structure of our *chromosomes*, the long stretches of DNA that contain our genes. Humans have 23 pairs of chromosomes, with one chromosome in each pair contributed by the mother and one by the father, for a total of 46 chromosomes. During the creation of new sperm and egg cells, the genetic material within each cell must be divided in half, so that each new cell receives only one chromosome from each pair.

The randomness of this process is what ultimately leads to the incredible diversity of human genetics. While each parent contributes an equal number of chromosomes, the specific chromosome from each parent that is transferred to the new cell is determined purely by chance. This means that no two sperm or eggs are exactly alike, and each new individual inherits a unique combination of genetic information from their parents. It also explains why siblings are similar yet different from each other.

Now let's consider how the process discovered by Mendel affects your own genetic makeup. Your father's DNA contained one set of chromosomes from your grandfather and ones from your grandmother. During the formation of your father's sperm, genetic information was passed down according to a randomized dynamic: each chromosome carried in your father's sperm could contain information from either your grandfather's or grandmother's DNA. Near-infinite combinations (some 8.3 million, to be precise, even before accounting for recombination) of our 23 individual chromosomes are possible, and every sperm or egg cell is different. When the egg is fertilized by the sperm, the paternal and maternal chromosomes combine, forming the unique chromosome pair combination of the new embryo.

The purpose of sexual breeding is exactly to generate diversity that will give rise to the corresponding variety in traits. This diversity is the engine of evolution, the generation of new combinations that may impart advantage for survival in a given environment.

But this brings us back to the mysterious dynamic first discovered in the garden of that Moravian abbey. Why are some genes dominant while others are recessive? In other words, why does one gene override the effects of the gene present on the matching chromosome?

To understand this better, let's delve into the molecular mechanism behind the mysterious reappearance of hidden traits in the third generation. The gene for seed shape produces a protein that controls the shape of the seed by modifying starch content. The starch also provides the plant with the energy required for germination.

In the case of Mendel's pea plants, the normal gene sequence contained the instructions for producing a functional enzyme that created the starched and rounded look of the smooth peas. The defective gene contained a mutation that disrupted this instruction, resulting in peas that were wrinkled. When both copies of the gene contained the mutation, the seed was unable to produce this functional enzyme, resulting in a wrinkled appearance. However, when one copy of the gene was normal, the seed was able to produce enough functional enzyme to modify starch and produce a round seed.

This example illustrates how a gene's sequence can affect the corresponding protein's structure and function, ultimately influencing an organism's physical traits. It also shows how, in any given individual, the presence or absence of a functional gene copy can determine whether a trait is dominant or recessive.

The situation can get more complicated when both copies are functional, giving rise to two different forms of a given protein. In this case, we encounter a situation of *codominance*, where the combination gives rise to a new feature in the progeny that did not exist in the parents. In Codominance, both alleles are expressed simultaneously, rather than one masking the other. Examples of codominance are flower color and human blood types.

Clinical genetic counseling relies heavily on the laws of recessive and dominant traits that Mendel established. These principles can help when a disorder is identified and we want to know both which gene is responsible for the disorder and from which parent it was inherited.

Working in the pediatric ward of any hospital is an intense experience, driven by the need to quickly diagnose and provide treatment to children in need while simultaneously caring for their worried parents. In the hospital in Jerusalem where I did my internship, things were made particularly complicated by the high rate of ultra-rare disorders, triggered by the high rate of consanguinity—marriage between close relatives—in some of the populations we served. This was the case with a patient whose name was Chanan.

Chanan's parents were cousins, who became a couple based on their society's belief that it was an obligation to marry within the family. These parents, Alina and Joseph, were both completely healthy, as were their first three children. When Chanan came along, he had severe medical challenges for which the parents were only partially prepared.

An ultrasound exam done during Alina's pregnancy had shown that their unborn child had a heart condition called *ventricular septal defect*. In 90% of newborns, this condition resolves itself spontaneously within a year. But Chanan was one of the less fortunate 10%. He also had other striking physical abnormalities. These included club feet, requiring treatment with casts, and enormous, widely separated eyes. The placement of his eyes reflected a defect in midline development—the process that forms the central structures of the face and body. Certain parts of the body that straddled an imaginary line separating the child's right and left sides developed abnormally, resulting in a huge fontanel (soft opening) in the front of his skull, a cleft tongue and palate, and a prominent coccyx at the base of his spine. Most dramatic, however, and most dangerous was Chanan's heart condition.

Usually, ventricular septal defects are characterized by a single hole between the two ventricles of the heart, two of the four chambers that, together, are the gatekeepers of the blood circulatory system. The left atrium receives oxygen-rich blood from the lungs, then passes it through the left ventricle of the heart. Then, the left ventricle pumps blood to the body. After delivering its oxygen, the depleted oxygen-poor blood reenters the heart through the right atrium, before the right ventricle pumps this blood back to the lungs. In Chanan's case, however, at least three holes in the wall between the ventricles disrupted this finely tuned dynamic.

Oxygenated blood arriving in Chanan's left ventricle leaked through the holes directly into the right ventricle. Not only did this reduce the amount of oxygenated blood delivered to the body, but it also sent some blood directly back to the lungs, together with blood that had completed its circuit through the body. This caused the heart to work less efficiently, making it work very hard to get the required blood into circulation. This is a dire condition that can end in death.

Because Chanan's heart had to work so hard, his circulatory system consumed all the energy the child received through food. He was fitted with a feeding tube, so that his heart would have enough energy to keep pumping and the rest of his body could continue to function. Our goal was to keep Chanan's overworked heart pumping until heart surgery could be performed.

I was watching over the child's preoperative treatment and studying the day-to-day medical decisions made by my wonderful clinical supervisor (remember, I was still an intern at the time), when I saw Dr. Eva Gross enter the ward with a triumphant smile. She carried a printed article from a medical journal that included a photo of a child treated a few years earlier by a different group. Based on the resemblance, which included a set of midline developmental defects, I might as well have been looking at a picture of my own patient, Chanan.

The diagnosis was reached: it was Frank-ter Haar syndrome, a rare condition that occurs when two abnormal copies of a recessive gene are inherited, one from the mother and the other from the father.

The geneticist on our ward enthusiastically took up the next challenge: to identify the specific abnormal gene that caused the disorder. The search began with an examination of

Alina and Joseph who, as close blood relations, were likely to share many similarities in their genes.

The geneticist's goal was to identify genes where each parent carried one silent, recessive mutation-mutations that, when inherited together, formed a matching pair in the child and caused disease. The sequencing (done as part of a worldwide collaboration) identified a candidate: a genetic mutation in a gene called *TKS4*, which was carried in one copy by Alina and Joseph and which appeared in its mutated form in the two copies they passed down to their son.

What is TKS4, and how did it trigger Chanan's extensive list of developmental defects? Proteins encoded by this gene function as a scaffold, bringing several proteins together so that they can carry out their roles in coordination. Without this TKS4 scaffold, the entire edifice falls, and the proteins cannot fulfill their various functions.

These proteins play a vital role in cell migration—a critical dynamic in embryonic development—because it enables cells that build specialized tissues to mature in the proper place. TKS4 coordinates the activity of two classes of proteins involved in this process. The first group includes enzymes that clear a path for the migrating cell. The second class of proteins provides the force within the cell that propels it forward.

Chanan's unusual features offered visible signs of a cell migration disorder. For example, his short stature was likely rooted in problems in bone formation, since normative bone growth relies on the active movement of cells that degrade and deposit bone material.

We knew that this syndrome is characterized by a range of developmental anomalies, including intellectual disability. We talked about this with the parents. But the first order of business was heart surgery. Happily, the operation went very well, closing the holes that had prevented normal cardiac function.

After remaining in the hospital for a few days for observation, Chanan was released, but not before a meeting was scheduled to discuss what lay ahead. Alina and Joseph already understood that their youngest child would face significant physical difficulties as well as intellectual challenges. Still, they insisted on emphasizing the positive, expressing their deep appreciation that the staff on the ward—the people who understood Chanan best—would continue to monitor the child in the years to come.

"Our baby is finally coming home, thanks to all of you," Alina said.

"We also feel blessed that Chanan is not our first child," Joseph added. "Now we know that the gene mutation we both carry creates a 25 percent risk for having another baby with Frank-ter Haar syndrome. Under the circumstances, we've decided that our family—two healthy parents, three healthy children, and all five of us available to love and care for Chanan—is exactly the right size."

Benny and Ehud later discussed the case over two more cups of coffee.

Benny: So do you think they might otherwise have aimed for a much larger family?

Ehud: Well, of course, there are no hard-and-fast rules, but in societies in which consanguineous marriage is accepted, a large number of children is not unusual. Then again, when parents have to contend with the challenge of raising a child with so many challenges, plans can fall by the wayside.

Benny: The number of tissue types affected by TKS4 mutations inherited from both parents is dramatic because it affects cellular mobility. From the earliest stages, every cell needs to migrate to a specific place. In the mature fetus, as in adults, mobility continues to play a role, like when developing nerves migrate to innervate target tissues or when white blood cells of the immune system change shape and squeeze between different tissue barriers. So as you know from your work in the clinic, treatment of this syndrome is extremely complex.

Ehud: Let me ask you, from your perspective as a scientist: If this gene is essential for so many different cell types, why isn't TKS4 mutation even more detrimental? For example, why was Chanan born with eyes that are very widely spaced, while other parts of his body look normal?

Benny: This question brings up the topic of backup systems. Just like an airplane has two or three backup systems for every critical instrument, in the body, many systems also have multiple mechanisms that can step up to perform the same task. It is possible that in some systems that require cell migration and movement, there are proteins other than TKS4 that can carry out the job.

Ehud: If that's the case, how is it possible to positively link developmental symptoms to one particular gene? How would you investigate the function of TKS4 in a particular type of cell, the cells that make up Chanan's cardiac tissue, for instance?

Benny: This is where animal models are extremely helpful. For ethical reasons, we can't conduct experiments on human tissues. But in laboratory experiments, we can mutate a gene of choice, say TKS4, and generate animal models that carry the mutant gene yet are healthy, because the mutated gene was inherited from only one parent. When these mice mate, a quarter of their progeny will be homozygous—that is, they'll have two copies of the mutated TKS4 gene, and the mutation will manifest in the phenotype. We can use these mice to follow embryonic and postnatal development. This allows us to learn a great deal about the mechanics that give rise to the human syndrome.

Ehud: But, as you've said, this genetic inheritance would affect many different types of cells and tissues. How would you study the mutation's effect in only one type of tissue, like the heart?

Benny: Genetic science provides tools for that as well. Today, we can not only mutate any gene of choice but also mutate it only in a distinct set of cells or in a specific tissue of choice. We can even define a time for the mutation to take hold, inactivating the protein. This allows us to get a great deal of information about the role a gene plays in a particular organ at each developmental stage without the added complication of defects in other tissues.

Ehud: This is impressive, but it doesn't provide a magic cure for complex cases like Chanan's.

Benny: That's true. But the more we understand the impact that recessive genes like TKS4 can have when passed down by both parents, the better prepared we are to limit health risk to the offspring. By the way, I assume you're not suggesting that if Alina and Joseph had been aware they both carried this gene, they would have chosen not to have children at all?

Ehud: Of course not! But with a one-in-four chance of having a sick child, genetic testing would give parents the option of controlling the outcome. For example, they could commit to conceiving through in vitro fertilization (IVF), something that would allow embryos to be tested for the mutation before being implanted in the womb. I'm not saying this is always appropriate, and frankly, Frank-ter Haar syndrome is so rare that it's not likely to be included in standard genetic testing anytime soon. But it's something to think about. As we learn more about how mutations in recessive genes increase the risk of developmental disorders, it may be possible to give parents a better chance to have what they want: a healthy child.

Chapter 5
Journey to the Center of the Cell

Now that we have met some patients and been introduced to the process of genetic counseling, we're ready for a deep dive into the molecular machinery that maintains and processes genetic information. Join us, if you will, on a trip to the nucleus, the cell's central command center.

Imagine yourself shrinking to a millionth of your original size, until you're small enough to enter a living cell. Once inside, you make your way toward the cell's center and pass through the membrane that surrounds the nucleus.

Look around. What you see could be described as a cross between a warehouse and a factory. First, the warehouse: all the genetic information that a cell needs to survive, and to create new copies of itself, is stored here in the nucleus, a space just one or two millionths of a meter in diameter.

Next, the factory: like a well-oiled machine, relevant bits of this genetic information are plucked out of storage, copied, and dispatched to one of many different production lines. What rolls off the production lines at the end of this process are proteins, the biological molecules that perform particular functions.

All this is impressive, but you also notice something else. The various production lines are adapting, producing different protein products for different cell types. Sometimes the rate the proteins are made at speeds up, and sometimes it slows down, often in a controlled manner.

All the individual bits of data that trigger this process are stored in *deoxyribonucleic acid*, *DNA*. Producing the proteins that make a fully functioning cell and organism is much more complicated than manufacturing an airplane, and timing is key. Just like you wouldn't want to install an airplane's engine before its wings are finished and fastened in place, the cell's genetic machinery can produce a finished product only if a great many things happen, in exactly the right order.

One of DNA's jobs is to carry the information that allows each cell to build up its components. DNA faithfully transmits the genetic code from one cell to another during cell division. Most important for our context, DNA is also responsible for transferring genetic information from one generation to the next on the level of the

E. Banne, B.-Z. Shilo, *Dialogue With Our Genes*,
https://doi.org/10.1007/978-981-95-9571-6_5

organism. Now that you've got the big picture, let's get down to the nuts and bolts of how all this happens.

DNA consists of four basic building blocks known as *bases* (also known as *nucleotides*). Bases link together to form long strands that are packed into *chromosomes*, each of which carries unique genetic information. The DNA molecule has a double-helix structure, where two complementary strands are paired together through specific chemical bonds that occur between the bases. Thymine (T) always pairs with adenine (A) on the opposite strand, while guanine (G) always pairs with cytosine (C). This pairing is crucial for the two main functions of DNA: accurate replication of genetic information and production of proteins.

Nonreproductive (somatic) human cells have 46 chromosomes, organized into 23 pairs. Reproductive cells—the egg and the sperm—each contain 23 chromosomes. When the sperm fertilizes an egg, the genetic information from both parents forms new chromosome pairs that determine the genetic inheritance passed down to the offspring. These chromosomes contain genes that determine physical traits such as eye color and height, making them crucial to our overall genetic makeup.

When a cell divides, it gives rise to two new cells, complete with a full set of genetic instructions. It's a beautiful system: the two paired strands of DNA separate from each other, like opening a zipper. Each free strand now consists of a string of bases—those A's, C's, G's, and T's–without their matching, preferred base pairs. This creates an attractive landing pad for free bases, who join up with the bases they prefer: T with A and G with C. Mediated by dedicated enzymes, this pairing up is the DNA replication that ensures that the two new cells get a full set of genetic instructions. It also ensures something incredible, considering that each one of us started our lives as a single fertilized egg cell: an almost endless number of reliable cell divisions that result in all the specialized cell types our bodies need.

Proteins carry out cellular function, following instructions contained in the DNA. But like a well-stocked factory, these instructions are stored in specific places and accessed by specific workers to ensure that proteins are generated correctly. The unit within the DNA that contains the information needed to produce a specific protein is called a *gene*. We humans have about 20,000 genes, located at specific points within every DNA molecule.

Now, information from the gene needs to be accessed, copied, and delivered to the protein production line. This is done by special molecules called *ribonucleic acid*, or *RNA* for short.

Unlike DNA, which takes the form of a double helix, RNA is a single-stranded molecule. Many copies of RNA can be produced from a single gene, and this is essentially how the rate of protein translation—the number of proteins produced based on the gene's instructions—can be controlled. RNA is produced in the nucleus. Then, through specialized pores in the nuclear membrane, it is delivered to the cytoplasm, where the information it contains is used to create proteins in specialized factories called *ribosomes*.

Proteins are built from building blocks called *amino acids*. There are 20 protein-building amino acids in humans; when connected to each other, they comprise the linear sequence of the protein. The information contained within the sequence of the

RNA is converted to the language of proteins. Conversion from the RNA language to the protein language is something like translating a text from one language to another.

Proteins initially take the form of linear sequences of amino acids. But then, proteins fold in intricate ways, forming specific three-dimensional structures that facilitate function. Ultimately, the structure of the protein will determine its activity.

Importantly, if there is a change in the DNA sequence of a particular gene, even on the level of a single base (nucleotide), this can alter the protein that the gene generates. Such a change may then give rise to pathological outcomes—like genetic disease—depending on the protein altered and the nature of the change.

5.1 How Scientists Read the Landscape

How do we know what happens along the journey from DNA to RNA to protein generation and, ultimately, to protein function? Today, scientists have laboratory technologies that allow us to synthesize and read DNA at will. Writing is done chemically, by harnessing the enzymes that help nucleotides line up in a specific order. In a complementary process, we can read the sequence of the DNA, using a method called *Sanger sequencing*, named after the British scientist Frederick Sanger.

Sanger was one of very few people to receive the Nobel Prize twice. His method, developed in 1977, was the most widely employed sequencing protocol for 40 years. Despite the emergence of faster and cheaper methods, it is still in use.

Sanger sequencing allows hundreds of bases to be read, making it possible to identify mutations of single nucleotides as well as deletions or duplications of small bits of genetic material. The method works by inserting chain-terminating nucleotide analogues into DNA during synthesis, each nucleotide labeled in a different fluorescent color. The nucleotide that appears in the final, terminal position on a synthesized chain generates fragments that can be separated by size, even if the DNA chains differ in length only by a single nucleotide. Separated through a gel using an electric field, these varying chain lengths create a visual ladder that can be interpreted to reveal the exact gene sequence.

Scientists rely on this experimental approach, and others, because it explains what happens in cells on the most fundamental level. Such basic research is frequently the first step toward techniques for preventing, treating, and possibly even curing genetic conditions.

Because scientific researchers, geneticists, and genetic counselors approach the same problems from a slightly different perspective, let's wander over to the hospital coffee shop and join Ehud and Benny for their ongoing conversation.

Ehud: Work in my clinic is so pressured, and the atmosphere is noisy. It makes me jealous to see how you and your team seem to work in an atmosphere that's downright relaxed!

Benny: I wouldn't say *relaxed*. It's in the nature of any ambitious research scientist to be driven, and we experience periods of great pressure and stress, especially when we find ourselves competing with other labs. But it does help to work in a quiet setting. That's what gives scientists the freedom to focus and to delve into the complicated problems we try to resolve.

Ehud: But how do the scientists select their problems? I know that for me, I never choose, because the problems come to me. Every new patient knocking at my door represents another puzzle to be solved.

Benny: Yes, in that sense, basic research is very different. It's been said that the mark of a great scientist is not in their skill in finding answers but in their ability to identify interesting questions. Here in the lab, we're not like an engineering or biotech company setting out to solve specific concrete challenges. Instead, we devote ourselves to open questions that trigger the curiosity of the person heading the lab. If the question is difficult enough and sufficiently intriguing, it could eventually lead to discoveries that spark similar research by other labs in the wider scientific community.

Ehud: In my practice, we grapple with the human stories that come out of the genes and molecular processes you study. Maybe next time we meet, you can see how we work with patients?

Benny: I'd love to.

Chapter 6
Seeing the Genetic Forest for the (Family) Trees

Clinical genetics—the medical discipline devoted to solving genetic mysteries—involves a partnership between two categories of people.

When a patient with a genetic condition first walks into a clinic, they are first examined by someone with a certain degree of expertise in dysmorphology. This is the fine and intuitive art of "guesstimating" the nature of a disorder based on the patient's physical and behavioral characteristics.

The second half of the partnership is populated by all the scientists who have contributed to the known list of genetic mutations that trigger specific disorders. This catalog, based on many years of work by scientific teams all over the world, makes it possible for physicians and clinical geneticists to focus their investigation on the small group of genetic mutations that might be the cause of a patient's unique characteristics.

But this way of describing the process of diagnosing genetic conditions leaves out an important part of the story: how genes and mutations became causatively linked to specific clinical disorders. Historically, the first step involved sleuthing out how genetic inheritance works in successive generations.

As you will recall, some of the earliest scientific studies of inheritance were performed by Gregor Mendel, a monk with a penchant for gardening. Mendel used successive generations of cross-pollinated pea plants to identify the ratio of plants that would inherit specific characteristics. Although he was unaware of the existence of genes (as the term had not yet been coined), Mendel intuitively understood that both "mother" and "father" plants pass down half of their hereditary traits to the next generation. He also realized that some of these traits were "dominant" while others were "recessive." The visible traits of the offspring plants appeared in the statistical ratios Mendel predicted, reflecting whether each plant carried dominant or recessive versions of a given gene.

Mendel's work was more or less forgotten and had little impact until the early twentieth century, when researchers revisited his findings. One of these was William Farabee, then a junior student of physical anthropology at Harvard. Farabee stepped

E. Banne, B.-Z. Shilo, *Dialogue With Our Genes*,
https://doi.org/10.1007/978-981-95-9571-6_6

out of obscurity to make what was then an astounding claim: that Mendelian genetics operate not just in plants but also in animals, including humans.

Mendel had suspected that the dynamics of inherited traits he observed in his pea patch might have a parallel in humans, but he did not pursue the matter. Farabee's dissertation, however, proved that the time was ripe for Mendelism on a broader scale. He wrote a treatise, titled "Heredity and Sexual Influences in Meristic Variation: A Study of Digital Malformations in Man," that tracked, over five generations, the collection of observable characteristics (or phenotypic pedigree) of a family affected by brachydactyly, a dominant genetic trait characterized by shortened fingers and stature. His study showed that the ratio of those with and without brachydactyly followed a pattern that could be explained by Mendel's ratios of dominant and recessive inheritance.

Later in his career, Farabee published findings about a human recessive trait, albinism, a condition marked by having little or no melanin pigment in the skin, hair, and eyes. By interviewing albinos encountered during his travels, Farabee confirmed that this trait also followed the three-to-one ratio in the second generation that Mendel had defined as typical of recessive genotypes.

Farabee's research resonates right down to today's approach to clinical genetics. Then as now, unlocking genetic mysteries begins with tracking a family's history over generations. The result of this interview is a disease-related family tree, also known as a *genetic pedigree*.

The family tree technique can provide powerful clues. For example, when a relatively common disease is caused by an inherited mutation in a single gene, assessment of the family history may lead to early diagnosis and more successful disease management.

Let's see how it works by going back to a typical day in the genetic counseling office.

Alicia Cook, a healthy young woman pregnant with her first child, was referred to me for counseling. Her brother had hemophilia, a life-threatening condition in which blood clotting is impaired, causing the sufferer to bleed severely from even a slight injury and, sometimes, even spontaneously. Alicia was concerned that she might be a silent carrier of the same condition and that the disease-causing gene might be passed down to her unborn child.

As Alicia described her family, I jotted down a family tree, using the symbols adopted by the clinical community for this purpose: squares for males, circles for females. In this graphical language, a blank shape indicates a healthy person. When the shape is filled in, it indicates the presence of the disease under investigation.

My diagram began with a blank circle linked to a blank square, representing Alicia and her partner, Rick, both of whom are healthy and have no signs of hemophilia:

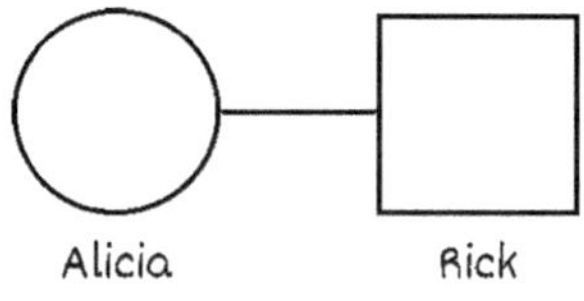

Rick's family had no known hemophiliacs, so I concentrated on Alicia's side of the family, filling in her siblings: her sister Trina and brother Alan, both healthy, and Brian, her brother with hemophilia. I indicated Brian's hemophilia by shading his square.

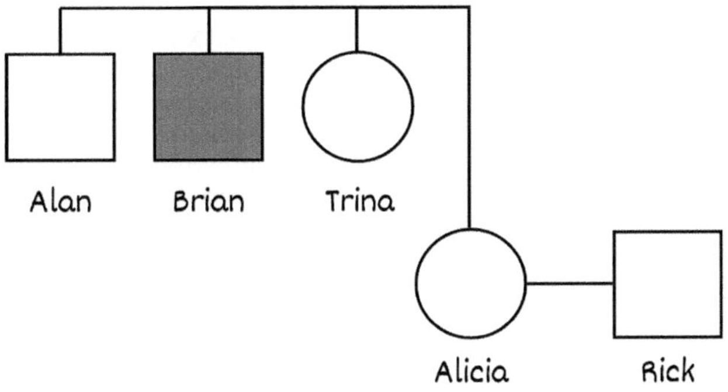

I then completed a three-generation picture, adding Alicia's parents, both healthy, and two sets of grandparents, including Alicia's maternal grandfather, a hemophiliac.

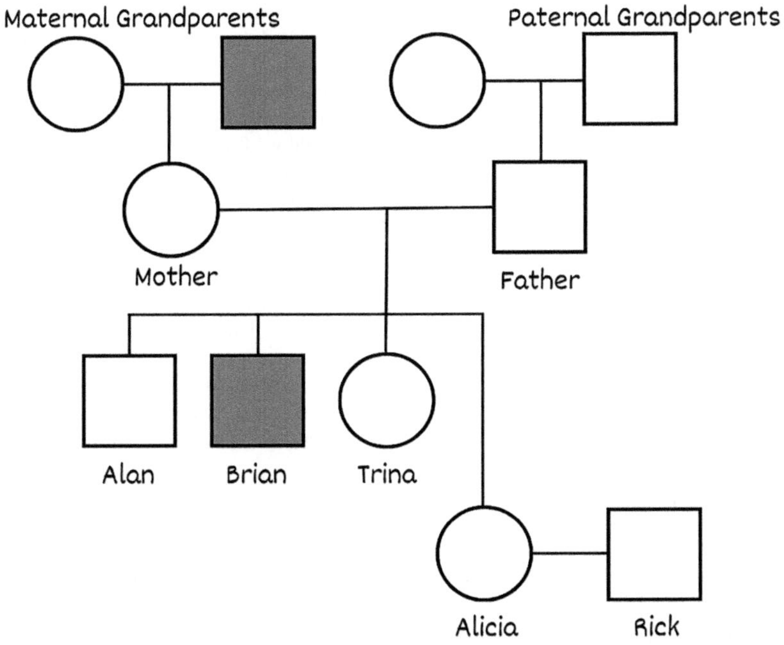

After completing this three-generation family tree, I was able to draw some conclusions, which I shared with Alicia.

"What we're looking at here is a family tree, showing how hemophilia has been passed down in three generations of your family," I explained. "Hemophilia is X-linked recessive. That means that the mutation is carried on the X chromosome and disease symptoms will only appear in offspring who lack a normal, nonmutated copy of the gene that causes hemophilia."

Alicia responded with a question that showed she had a good level of basic knowledge about genetics.

"But women have two X chromosomes," she said. "Why does it look like only men in my family have this disease? Wouldn't the extra X give women an even greater chance of getting it?"

"Excellent question," I responded. "Actually, that second X is what helps most women—but not all—to avoid hemophilia. Remember when I said this condition is recessive? That means that if women receive a mixed inheritance consisting of one healthy copy of the gene and one copy that's mutated—something that can't happen to men because they only get one X chromosome—the dominant, healthy copy will override the recessive trait. The mutation will still be part of her genetic makeup, and we therefore relate to her as a *carrier*, but symptoms will not appear in her phenotype."

"So what you're saying is that men don't have the back-up gene that gives them a second chance," she said. "Once they inherit that single gene, on their single X chromosome, hemophilia is inevitable."

"Exactly," I said. "Hemophilia is rare in women because both X chromosomes have to carry the mutation or one copy of the gene has to be missing or nonfunctioning."

Alicia still seemed a bit confused. Using the family tree, I pointed out the progress of the disease through the generations.

"When it comes to inheritance, it's a roll of the dice," I replied. "Your maternal grandfather clearly carried the hemophilia gene, but when your mother's genetic material was passed down to her, this inheritance was silenced—most likely by the dominant, normal form of the gene carried by your grandmother. She still had the mutated gene to pass down to the next generation, but she didn't have the symptoms."

Alicia was quiet for a moment. Then she said, "I can see how this family tree helps map out the statistical chance of my baby being born with this disease," she said. "But we don't know if we're having a boy or a girl at this point. Should we find out? Will it make any difference? How might this disease affect my baby, and is there anything we can do to help?"

I explained that there was prenatal testing available that would allow Alicia and her partner to know for certain whether the child they were expecting had hemophilia as part of their genetic inheritance. They could also select pregestational molecular testing to choose an embryo free from the disease. I told her that recent biomedical advances had raised hopes for successful intervention via gene therapy, which seeks to block the disease by altering a child's genetic profile. However, there was insufficient data to know whether such an approach could lead to a long-term cure.

"But it's important to recognize that medical intervention can allow a child born with hemophilia to live a long and healthy life," I said. "The main treatment for hemophilia is based on infusions that replace the blood clotting factor that is missing or low in hemophiliacs' blood."

As I traced hemophilia through Alicia's family tree, I couldn't help but think of a famous case study familiar to every genetic counselor.

Queen Victoria, who ruled Britain for most of the nineteenth century, is believed to have been a carrier of hemophilia B, a relatively rare form of the disease that is associated with a deficiency in a blood protein called *factor IX*, which promotes clotting. She passed the gene for this trait on to Princesses Alice and Beatrice—two of her five daughters—and passed the disease to her youngest son, Prince Leopold, who died in his early 30s. The prince's three older brothers were unaffected by the disease.

The daughters' hidden genetic inheritance made a grand entrance in the next two generations. Alice and Beatrice married appropriately high-born spouses, transmitting the genetic mutation to children and grandchildren born to the royal houses of Spain, Germany, and Russia.

Queen Victoria's rein of almost 64 years, longer than any of her royal predecessors, was marked by important scientific advances, as well as Britain's military and economic expansion. It is ironic, therefore, that while Victoria was presiding over the events that established her as the dominant leader of her era, her own family was dominated by a hidden genetic abnormality. If, rather than activating her armies for conquest, Queen Victoria had set her sights on the neighboring Austrian empire, where an unassuming monk was quietly examining pea plants, she might have helped bring the principles of Mendelian inheritance to light earlier than they were ultimately understood.

Unfortunately, it was not until 1900, the last year of Victoria's reign, that three botanists—Hugo de Vries, Carl Correns, and Erich von Tschermak—independently rediscovered Mendel's findings, a full generation after Mendel's papers were published. William Farabee then used these scientists' work to hit the ground running in his own studies of human inheritance.

The next time I visited Benny's lab, I described how this clinical consultation reminded me of Queen Victoria and her famous hemophiliac descendants.

"Maybe we should honor Her Majesty by brewing a nice pot of tea, preferably with some milk," Benny said with a smile.

"Too much work," I replied, helping myself to my usual instant coffee.

"Here's to our common struggle," Benny said, toasting me with his mug. "May we both find the genes we seek."

"And may we find the genes we didn't know to look for," I added. "That's one of the exciting things. When a new patient walks into my office with symptoms that don't obviously line up with any known syndrome, I know I could be witnessing the phenotype-level expression of a genetic abnormality that has never been isolated or fully understood. The initial intake exam—and particularly the creation of a family tree—establishes the terms of investigation."

"Sure," Benny agreed. "Establishing the pattern of inheritance—on the most basic level, whether it is dominant or recessive—clarifies expectations about how symptoms are likely to appear. Without the pedigree, all our tests, from karyotyping to next-generation sequencing, would not help us link an unexpected genetic mutation to a patient's unique characteristics. In the case of known mutations, family trees provide an important first hint. And you're right, they are especially instrumental when trying to identify a disease gene that has never been described before."

I completed Benny's thought. "And the only way to be sure we've made the correct link is to look at how the symptoms appear, or don't, in other family members. If the condition is known to be caused by a dominant mutation, then healthy family members, those without symptoms, can be assumed to lack the mutated gene because one copy, or two, would cause symptoms to appear. When the genetic mutation is recessive, things are a little more complicated. Family members lacking symptoms may carry the abnormal gene as a single copy, hidden by the normal copy."

"But the functional impact of the genes we trace through family trees can remain an unresolved mystery," Benny said. "And good thing, too. Otherwise, I might be out of a job. My laboratory research focuses specifically on how genes drive development and determine function."

"Don't worry," I said. "Genetic counselors will always rely on findings from basic science labs. Your studies in cells and animal models are incredibly important because it's ethically impossible to do such experiments on humans. Scientists' experiments allow us to see the impact of genetic changes, something that can reveal the cause of the symptoms we observe in human patients."

"Happy to be of service," Benny said, gesturing with his coffee mug to a lab surface, where half a dozen flasks were buzzing with fruit flies. "On behalf of me and my model, you're welcome."

Chapter 7
What Flies and Worms Teach Us About Human Disease

Not long ago, I found myself left holding the bag. My wife, who loves shopping for clothes, had discovered a family-run shop and was in the dressing room, happily trying on outfit after outfit. Based on years of married experience, I settled in to wait it out, chatting with the store's owner, a 30-something hipster whose own distinctive attire indicated he might be the designer behind the store's colorful merchandise.

The owner, who introduced himself as Jimmy, quickly disabused me of this notion. "Actually, my wife is the one with the eye for fashion," he said. "I handle orders through our website and manage the store, which gives me plenty of time to chat with customers. Can I make you some coffee while you wait?"

Sure enough, we had plenty of time to get to know each other. When Jimmy found out I was both a pediatrician and a geneticist, his face and tone turned serious.

"I hope you don't mind me asking you a medical question on your day off," he said, "but we're worried about our daughter. She's always had trouble eating, and she's not growing well. Our pediatrician sent us to specialists. All they say is they think our daughter has some kind of genetic disease. And now you walk into the store." Jimmy's voice trailed off, clearly uncomfortable to be asking for help.

"I would be happy to see your daughter," I said. "How soon can you come?"

Jimmy, his wife Shauna, and their daughter Kate came to my clinic the next day. Kate was seven, but with her short stature and tiny physical features, she could easily be mistaken for a five-year-old. The child had additional abnormalities that indicated the presence of a genetic syndrome: her eyes leaned at a lateral, downward angle, and she had a relatively short neck. Her most distinctive physical trait, however, was webbing. This is an extension of the skin that widens the posterior neck. It is as though the skin from Kate's tiny shoulders decided to connect directly to the base of her skull.

I was not surprised to learn that Kate had had feeding problems since early childhood and had experienced developmental delays.

"We had to install a feeding tube when she was a toddler," Shauna said, the slight tremor in her voice betraying the emotional impact of that trauma. "And she didn't start talking until she was three. Her special education teachers were wonderful. But she's in second grade now and still doesn't recognize all her letters."

Jimmy was quick to point out that, as parents, they had done everything they could. "Shauna has been Kate's champion from day one," he said. "She hired private tutors and takes Kate to after-school activities while I'm minding the store. Our goal is not to make Kate catch up with her peers—we understand that may never happen. Our wish is to make Kate feel loved for the person she is."

E. Banne, B.-Z. Shilo, *Dialogue With Our Genes*,
https://doi.org/10.1007/978-981-95-9571-6_7

I smiled, silently admiring the young couple's attitude and dedication. However, I felt a *but* coming. Sure enough, it came from Shauna.

"I'm 38 years old, and it's time for Jimmy and me to be thinking about a second child," she said. "I love Kate more than life itself, but I can't go through this again. Can genetic testing help ensure that our next pregnancy results in a healthy child?"

Jimmy and Shauna chose to explore the option of genetic testing. Kate's collection of symptoms indicated to me that she probably had Noonan syndrome, a condition linked to mutations in a genetic network called the *RAS pathway*. The results of their test proved that my hunch was right.

Imagine the RAS pathway as a game of telephone, where a message is whispered from one person to the next. A message passed between proteins in the RAS pathway tells the cell to divide.

But what if one protein in the network is a bit too enthusiastic, yelling the message instead of whispering it? This is what happens in Noonan syndrome: a mutation causes signaling to become hyperactive, driving an uptick in signals that can lead to cells dividing uncontrollably, a hallmark of cancer.

No gene or protein functions in isolation, and the individual proteins in the RAS pathway usually work together to keep cell division at normal levels. But what happens when one or more of the RAS pathway genes are mutated? This can have a serious impact on human health: more than 25% of human tumors carry mutations in RAS genes. These mutations can lead to cell division under situations where the cells would normally not divide.

Where can one get an even broader perspective on the RAS gene network? Astoundingly, one of the scientists whose original research led to our ability to investigate the inner workings of the RAS pathway was my patient Shiri's grandfather, Professor Ben-Zion Shilo.

Professor Ben-Zion Shilo—Benny to his friends, including me—is a faculty member in the Department of Molecular Genetics at the Weizmann Institute of Science, a postgraduate research institution located in Rehovot, Israel.

Benny was responsible for some of the most important basic research related to the RAS pathway. Born in Jerusalem, Benny moved to Boston in the late 1970s to become a postdoctoral fellow at the Massachusetts Institute of Technology, under the direction of Professor Robert Weinberg. Robert—Bob—was still a young scientist at the time, but he would soon be recognized as a world authority on cancer.

In a series of seminal experiments, the Weinberg team showed that by transferring DNA from cancer cells to normal cells, cancerous properties can be triggered in cells that previously showed no sign of malignancy. These investigations paved the way to the identification and isolation of the first oncogenes—genes that cause cancer—in humans.

Bob Weinberg's experiments, in which my friend Benny participated, revealed the involvement of RAS genes in cancer onset. But for Benny, this raised additional questions: What is the function of RAS genes in healthy, noncancerous cells? Moreover, how broadly are RAS genes found, and how do they operate elsewhere in the animal kingdom?

Benny's PhD research, at the Hebrew University of Jerusalem, focused on the genetics of baker's yeast, the same type of yeast used for making beer, wine, and bread. The organization of yeast is similar to our own cells, making them an ideal model system for research aimed at examining human biological processes. Benny set out to determine if there were RAS genes in yeast. If so, this would open endless possibilities for the study of basic processes that occur in humans—like cell growth and replication—using yeast as a model system.

During this period, far before the sequencing of the human genome completed in 2001, there was no obvious way to determine whether a gene from one species was also present in another. However, Benny was young and undeterred.

He recalls cornering Bob at the lab Christmas party in 1980 and throwing down the gauntlet by making an unsubstantiated declaration. "Today," he said, "I started the experiments that will reveal the normal function of cancer-causing genes."

With the help of his colleagues, Benny collected DNA from three different species: yeast, worms, and flies. Using the cancer-causing RAS-related DNA isolated in the Weinberg lab, Benny produced a radioactive probe that would emit a signal when the RAS-related genes he was looking for created a chemical bond.

Lo and behold, Benny found the RAS genes in the genome of the fruit fly. In fact, in terms of amino-acid sequence—the order in which the building blocks of the protein appeared—RAS produced in flies was eventually shown to be 75% identical to the human protein. Later, using similar techniques, other labs found RAS sequences in worms and in baker's yeast.

Benny's discoveries demonstrated something extremely fundamental: that the RAS gene is present in all multicellular organisms. This proves that RAS is an ancient gene that existed in the common ancestor of flies and man, at least 600 million years ago, and existed even earlier, when yeast and other single-celled organisms emerged. The conservation of the RAS protein throughout evolution indicates that it is essential, not just for the function of individual cells but also for the survival of multicellular organisms down to the present day.

Most importantly, this discovery opened a window through which we can view the function of the RAS gene in its normal, non-disease related context. Using the rich genetic toolkit available for the scientific investigation of biological processes in worms and flies, scientists have been able to characterize the entire cascade driven by proteins comprising the RAS pathway and identify other genes that interact with RAS.

Extensive genetic work in flies and worms, performed by Benny and scientists working in many other labs, has demonstrated that RAS is a central hub for transmission of information through the cellular membrane. In addition to its involvement in the abnormal signaling that can contribute to cancer, the pathway has also been proven to play an essential role in embryonic development. It is involved in forming, for example, the elaborate structure of the fruit fly's compound eye. By studying these dynamics in flies, we gain new insights into RAS-related dynamics in human cells that occur both in normal development and health, and in disease.

Many people may find it perplexing, even disturbing, to learn that we are similar in many ways to flies and worms. However, for scientists, our common genetic heritage provides a unifying view of biology across species. This drives our capacity to use discoveries attained in animals distantly related to us to advance medical diagnosis and treatment of human patients. In cases where the impact of a particular mutated and disease-related gene remains elusive, research scientists can test the activity of that gene in a model organism to help identify the gene's possible function.

Kate's diagnosis was fresh in my mind when Omar, a new patient, stepped into my office. Even though this fully grown man looked nothing like Jimmy and Shauna's diminutive daughter, his collection of symptoms made it obvious: like Kate, I believed that Omar had Noonan syndrome.

Omar and his wife Nasreen were referred to me by our hospital's carrier-testing department. Carrier testing is a procedure used to determine if a person has a recessive genetic condition, caused by abnormal traits that, while not expressed in the parents, can be passed down to unborn offspring. As the nurse immediately noticed, Omar had an atypical appearance, indicating that specialized testing might be in order.

As I greeted the young couple, I took note of Omar's unusual features. He was short, with a widened neck and drooping eyebrows. As Omar summarized his medical history, additional hints emerged, including a childhood procedure to repair his cleft lip and palate and, later in life, two open-chest cardiac surgeries.

In response to my questions about her family, Nasreen was somewhat shy, but Omar, a boisterously confident young man of 27, had no such inhibitions.

"I warn you, if you want to fit my family tree on one sheet of paper, you'd better write small," he joked. Omar's long list of brothers, sisters, aunts, and uncles was indeed a challenge to summarize, but Omar told me not to worry about details. "The most important thing to remember about my family," he said with a grin, "is that my mother says that I am the handsomest."

A physical examination revealed additional physical characteristics including short fingers, a low frontal hairline, and a small jaw that caused Omar's teeth to be severely crowded together.

"I understand you two came here to do standard carrier testing, but Omar has some traits that I suspect may be linked to a genetic syndrome," I said. "Although this hasn't prevented him from living a full life, such a syndrome could be passed down and could play out differently in any children you have. That's why, in addition to Nasreen's standard test, I want to order a specialized genetic screening that's done outside of the hospital. We'll know more in a couple weeks."

Immediately, Omar's humor reflex kicked in. "I don't know, Doctor, my mother says she wants a new grandchild before next Ramadan," he joked.

With a sharp glance at her husband, Nasreen made her feelings clear. "Omar, you know I love you, but I think we should wait until we finish your genetic work-up," she said. "We need this information."

Slightly abashed, Omar turned to his wife and then gently took her hand. "Inshallah," he said, quietly.

When couples who want to have children come to our clinic, we advise carrier screening to test for the possibility that both members of the couple might carry a disease-causing mutation in the same gene. This kind of testing is usually done to identify recessive disorders that will only appear in offspring when both copies of the gene is mutated. If both parents are asymptomatic but each carry one recessive mutation, there is a 25% chance that their child will be born with the observable characteristics, or phenotype, of the disease.

On the other hand, some genetic diseases are dominant, meaning that only one mutated gene is necessary for disease symptoms to be expressed. In this scenario, there is a 50% chance that a disease-causing mutation will be passed down to the next generation. The stakes are high in this numbers game, making it important for Omar and Nasreen to finish their genetic workup before thinking about getting pregnant.

We sent Omar's DNA sample to the same gene panel where we sent Kate's. These gene panels examine a set of genes that comprise a signal pathway, as in the case of the RAS pathway. This pathway is actually a collection of communicating proteins within the cell that deliver a signal from the outside of the cell and into the nucleus, where it influences which genes will be transcribed. Mutations in each of these genes can cause the same symptoms seen in Noonan syndrome.

When the results came back, they showed that Omar, like Kate, had an activating mutation in the SOS1 gene. The SOS protein is normally responsible for activating the RAS protein network. The abnormal function of the SOS protein explains why both 7-year-old Kate and 27-year-old Omar suffered from Noonan syndrome.

The medical term for abnormal mutations that affect the RAS pathway is *RASopathies*. They are a significant group of genetic syndromes. While each specific RASopathy is rare, as a group, mutations in the genes that code for individual components of the RAS pathway cause many of the learning and developmental problems with a genetic cause. Since RASopathy patients share so many features, it is easier to look at them collectively as a group than to divide them. Mutations in many different genes can produce, in the end, the same result.

It is thanks to the work of scientists like Benny that I was able to examine my patients, like Kate and Omar, and make an educated guess about the syndrome that caused their developmental symptoms. And that guess—verified with testing and then responded to appropriately by supportive family and friends—marks the first step toward a life that can be lived to its fullest.

Ehud: You know, Benny, the discoveries you've made during your career as a scientist have had such a clear impact on my work in genetics that—though I hate to admit it—I probably owe you more than a cup of coffee.

Benny: I know of an excellent restaurant.

Ehud: Seriously, you might be surprised how disturbing it can be to some patients to realize that what we know about their genes was first observed in fruit flies or worms.

Benny: It's true, maybe not every patient needs to delve into the science. But all patients benefit from what scientists have learned from animals. In medical genetics specifically, I think that's going to be even more true in the future.

Ehud: That's good to hear. There are so many cases in which, even after genetic analysis and sequencing, we can't link a particular gene to a patient's pathology. The fact that several genes may be altered only makes the puzzle harder to solve.

Benny. You're right. When I discovered the conservation of cancer-causing genes in flies all those years ago, it was a revelation that altered our way of thinking. But now, some estimates say that 6000 human genes may be associated with diseases. This makes identifying the gene-to-disease connection very complicated. Since an astounding 75% of genes in the Drosophila fruit fly have parallels in the human genome—that's about 10,000 fruit fly genes!—animal models are an excellent place to start.

Ehud: But how can this genome-wide similarity help diagnose genetic diseases in humans? One of the frustrating things is that even if we can identify genetic variants, we can't necessarily tell if the variants we find are the cause of a particular medical problem or an innocent bystander.

Benny: This is exactly where flies may help. Remember that many of the fly genes have structurally similar counterparts in humans? To find out if they are similar in function, not just in structure, we ask, can a human gene replace the fly gene? It turns out that in more than 50% of the cases, the human gene is functional when integrated into the fruit fly genome. This opens enormous possibilities to enhance diagnosis of human diseases.

Ehud: Great. All my patients need to hear is that mad scientists are putting human genes into flies. They'll go running for the door.

Benny: Actually, I think that some of your patients may find this very interesting and not scary at all. Of course, it's necessary to emphasize that scientists are not creating some sort of Frankenstein monster here. The human genes transferred are those we suspect of being associated with some very specific function or developmental process. Because the fly life cycle is so short, we can see the results very quickly and get definitive answers. This makes fruit fly genetics a very cost-effective platform for solving genetic problems.

But let's go back to your question. How do we identify disease-causing genes? Let's say you have a patient in which DNA sequencing has identified variants in two different genes. You want to know which of these variants is causing the syndrome. If there are *homologues*—parallel genes, something that has to be confirmed by experiment—you can then transfer the human gene in its mutated form. This creates a yes-or-no scenario. If the fly develops normally, the suspected gene is innocent. But if the mutated gene causes defective function or development in the fly, you know that this mutation is involved in causing the human genetic syndrome.

Ehud: And you say "involved in causing" because the cause might be the interaction of more than one mutated gene.

Benny: Correct.

Ehud: So genetically altered fruit flies can be seen as test tubes for identifying the involvement of specific genes in disease. How about the next step? Do you think that flies will be able to reveal the molecular cause of a medical problem?

Benny: Absolutely. Even though on the level of the organism, fruit flies and humans look very different, there are some profound similarities. For example, the development of the nervous system, the heart, and the muscles is driven by similar sets of genes in both flies and humans. This means that using the fruit fly model, we can get very detailed, molecular information about what went wrong in human patients who have syndromes that affect these parts of the body.

Ehud: Interesting, but it sounds like water under the bridge when what most of my patients want—especially the parents of young children with genetic syndromes—is hope.

Benny: Actually, I think there is hope. Let's say a patient's problems are related to a genetic variant that causes over- or underproduction of a particular protein. Studying this variant in an animal model may provide a platform for testing drugs that might reduce or enhance the activity of that protein. Such a drug—produced *on the fly*, so to speak—may not cure the problem. It certainly can't turn back the clock to give a child born with a genetic syndrome a second chance to develop in a way that's completely normative. But targeted drug development in animal platforms can generate new treatments that may give such children a better chance for attaining a higher level of function and a richer, more fulfilling life.

Chapter 8
Chromosomes: Packing for the Genetic Journey

From the moment that Mrs. Strauss and her daughter Gitti stepped into my office, I saw that the child—actually, a short and very childlike teenager—had a recessed chin and a back hump, wore hearing aids, and had signs of a severe intellectual disability. As a physician and clinical geneticist, it is not unusual for me to see and treat such patients. But when the mother presented me with the referral from Gitti's pediatrician, I was taken aback.

"Your doctor writes that, at birth, your daughter was diagnosed with trisomy 18," I said to Mrs. Strauss, whose extremely modest clothing identified her as a member of the Hassidic ultraorthodox community. "Are you familiar with this condition?"

"I read about trisomy 18 as soon as we got the diagnosis," she answered. "But it didn't seem to match Gitti's symptoms. For years, I believed that the doctors made a mistake, but I've never been sure."

"Your intuition was right on target," I responded. "As you perhaps remember from your reading, most children born with trisomy 18 do not survive. Although your daughter has certain disabilities, the fact that she is in her teens and seems to be in good health indicates that the diagnosis was probably incorrect. What I suggest is a blood test that will hopefully help us identify the root of the problem."

Mrs. Strauss had no documentation about the genetic testing performed soon after her daughter's birth, but Dr. Banne knew the diagnosis was wrong. In trisomy 18, people are born with three rather than two copies of the 18th chromosome. Most children do not survives early childhood, much less long enough to walk into a doctor's office at 15 years of age. So it was necessary to start from scratch. The first step was to clarify Gitti's genetic profile by examining her chromosomes. But to fully understand Gitti's story, we must pause here for another clarification: what are chromosomes anyway?

Chromosomes are structured units of DNA located within the cell nucleus that play a vital role in transmitting hereditary traits from one generation to the next. Each chromosome carries genes that determine specific traits, such as the presence of sickle-shaped red-blood cells or the development of cystic fibrosis, a congenital lung disease. Chromosomes come in pairs, one version inherited from the mother and the other from the father. The organization of chromosomes and the dynamic changes they undergo control the expression of genetic traits and shape our unique identities. Ultimately, chromosomes play a central role in shaping who we are.

E. Banne, B.-Z. Shilo, *Dialogue With Our Genes*,
https://doi.org/10.1007/978-981-95-9571-6_8

But not only us. Chromosomes serve as the unique identifier for all living organisms.

Humans typically have 46 chromosomes, arranged in 23 pairs made up of complementary partner genes inherited from the father on one side and the mother on the other (more about the significance of this pairing later).

Elephants have 56 chromosomes, and corn has 20 chromosomes, but this is not because elephants are bigger than we are and ears of corn are smaller. In fact, the diminutive hermit crab holds the record for the most chromosomes in the animal kingdom, with an impressive 254 chromosomes. This amount simply reflects the fact that the hermit crab genome—its complete repertoire of genes—is organized into a relatively large number of small, discrete collections of genetic material.

Ironically, despite the fact chromosomes hold our genetic material in a very specific and organized way, for most of their life cycle, chromosomes hardly look organized at all.

Most of us are familiar with the double helix structure of DNA. This twisted ladder shape was first revealed as the organizing principle of the DNA molecule through the pioneering work of British chemist and X-ray crystallographer Rosalind Franklin. Later work by biologist James Watson and physicist Francis Crick earned those two scientists and Maurice Wilkins—and not Franklin, a point of historical contention—the 1962 Nobel Prize for discovering DNA's molecular structure.

The distinctive double helix of DNA looks solid enough when set out on the pages of a textbook. In real life, however, DNA molecules are fabulously flexible. They are as floppy as individual strands of spaghetti. How then, does our genetic material stay organized and reliably pass down genetic inheritance from parents to offspring?

Except for a brief period during cell division, our 46 chromosomes exist as separate entities within the nucleus, entirely unconnected to their complementary partner chromosomes. At first glance, the multistranded bowl of cellular spaghetti may seem chaotic. However, closer examination reveals DNA's remarkable organization and regulation, a machine for reading and executing genetic instructions.

And given the complexity of the task, a machine is indeed what's needed. DNA has an incredibly intricate molecular structure, consisting in humans of three billion base pairs. As noted before, each pair is built from one of four nitrogen-containing bases (or nucleotides): adenine (A), cytosine (C), guanine (G), and thymine (T). The bases pair off according to specific rules, with A linking to T, and G linking to C. This vast but ultimately orderly arrangement is what encodes the genetic information found in each cell. It determines how genes are expressed and which traits appear.

If these base pairs from a single nucleus were lined up end-to-end, they would extend over a meter. This vast amount of data, packaged into structural units called *chromosomes*, needs to come together in an organized way so it can be replicated, a process ensuring that a cell can pass down a complete set of genes to the two daughter cells produced during cell division.

Our genetic material must pull off this feat of replication in a small, constricted space; all our DNA is compactly stored within a nucleus measuring only a few millionths of a meter across. This physical compression of complex genetic data can be likened to packing a room-sized computer's worth of information into a small thumb drive.

How does DNA manage to store such a vast amount of information within a small space and ensure the accurate transfer of genetic information from cell to cell, organism to organism, and generation to generation? Its success depends on a skill that should be mastered by every frequent flyer: smart packing.

8.1 Packing the Nucleus

A smart packer knows how to organize things into a small space, while keeping certain items easily accessible. This is the strategy that characterizes how chromosomes are packed inside the nucleus between cell divisions.

A critical player in this molecular packing scheme is the histone protein, which acts like a spool around which the DNA strands that form individual chromosomes are wrapped. Eight histones wrapped with DNA form another organizational unit called a *nucleosome*. Multiple nucleosomes are linked together like beads on a necklace, repeatedly twisting and turning.

The nucleosome necklace resembles a repeatedly twisted rubber band that buckles and doubles up upon itself. As a result, this collection of genetic information becomes shorter, wider, and extremely compact. This structure also protects these long, narrow DNA molecules from shearing forces that can cause damage.

The way that DNA strands wrap around histones and form nucleosomes plays a critical role in regulating gene expression. Normal cellular function depends on specific genes being activated at specific times. It also depends on other dormant genes remaining silenced—that is, nonactivated. But what determines which genes get a wake-up call?

The twisted, doubled-up structure of the nucleosome necklace keeps certain genes under wraps, far from the surface, where they remain inactive and unexpressed. At the same time, the twisting places other crucial genes on the nucleosome necklace's surface. This makes them readily available for gene expression and protein synthesis. In this way, DNA's organizational structure ensures that the right genes are activated at the right time, allowing for proper cellular function and maintaining the integrity of genetic information passed down to the next generation during cellular replication.

This DNA packing procedure, which ensures that some genes are expressed while others are not, works … most of the time. When it doesn't, a genetic counselor may be called in to determine what happened. This is why Gitti and her mother were sent to Dr. Banne, to see whether the girl's developmental problems were rooted in abnormalities related to chromosome structure.

Let's listen in to their conversation at the genetic counseling clinic.

"Gitti is our blessing, the youngest of our 10, and my other children have always looked out for her, both because of her special needs and because we all love her," said Mrs. Strauss, offering a tight smile that betrayed both her family pride and her concern. "I admit I was surprised to be referred for genetic counseling. After all, when I was pregnant, my doctor warned me that the baby would never survive, but obviously, here we are."

"It can be appropriate to consult with a geneticist at any stage of a child's development, and even in adulthood," I assured her. "Despite her developmental limitations, your daughter is clearly well cared for and is thriving. Still, it's worth investigating what caused her symptoms, so we'll take a look together."

Examining the child's referral, I noted that Gitti was previously diagnosed with a syndrome that stems from a significant visible abnormality in the 18th chromosome—something that in no way matched the child's health status or level of functioning. Turning to Mrs. Strauss, I told her, "If you will authorize a blood test, we can do a lab procedure called *karyotyping*. Karyotyping will provide a picture of Gitti's chromosomes. It will hopefully give us a more precise picture of your daughter's genetic profile and make sure she is getting the best possible medical care."

Mrs. Strauss was silent for a moment. "Doctor, I'm sure you're aware that some members of our Orthodox Jewish community choose not to perform genetic tests before a child is born. We fear this type of screening might scare parents into giving up on a soul created in God's image just because the child may look a little different. Well, Gitti's been part of our family for 15 years, and we will never give up on her. We want to understand her medical condition. You have my permission to do the test."

8.2 Repacking for Cell Division

Remember our discussion of smart packing? When cells divide, their genetic material is reorganized for easy transfer to the next generation.

When arriving at the hotel for a business trip, it is best to have everything packed tightly in a suitcase. But if you want to get to your various appointments on time, reorganizing your clothes ensures that what you need is easily accessible. Let's see how this works on the cellular level, with genetic information serving as the clothing that needs to be moved efficiently when one mother cell divides, creating two new daughter cells.

The two nucleotides that sit on opposite strands of the DNA double helix and form chemical bonds with one another are called *base pairs*. The twisting that allows three billion base pairs of every strand of DNA to be packed tightly into the cellular nucleus is an impressive feat of biomolecular engineering. But for a genetic counselor interested in finding out what's happening on a particular chromosome—say, the 18th chromosome under suspicion as the source of Gitti's developmental abnormalities—this seemingly impenetrable jumble of chromosomes puts critical aspects of the genetic picture tantalizingly out of reach.

Until it doesn't. During cell division, genetic material in the nucleus undergoes a radical change. The tightly packed chromosomes are released from the jumble and reorganized, readying themselves to be distributed into two newly formed daughter cells. This dynamic is as complex as it is critical, because it ensures the accurate transmission of all the cell's hereditary information to the next generation.

To pass on a complete set of genes to each daughter cell, our DNA must create a copy of itself. This copying, occurring immediately prior to cell division, doubles the amount of genetic material in the nucleus. As the nucleus was already tightly packed, you might think it would result in chromosomal chaos. But thanks to billions of years of evolution, Nature has everything under control.

When cell division begins, the chromosomes perform a tightly choreographed dance. They align themselves in the center of the nucleus, with each chromosome containing both the original and the duplicated copy, loosely attached to each other at one point. A pinch down the center of each chromosome divides these duplicated sets of chromosomes and pulls them in opposite directions, ensuring an equitable distribution of genetic assets between the newly formed daughter cells.

This division of genetic material during cell division usually works. But unfortunately, it is not foolproof.

Sometimes when the cell divides, two copies of a particular chromosome, rather than just one, are delivered to one of the daughter cells. When this error occurs during the formation of cells that drive fertilization—sperm for the male, egg for the female—the results can be disastrous. In such a case, even though no abnormality is discernible in the mother or father, a chromosomal syndrome will be passed down to their children. This is the basis of trisomy 18. It is also the dynamic behind additional genetic conditions including Down syndrome, which occurs when a child has three copies, instead of two, of the 21st chromosome.

So what happened in Gitti's case? This is where karyotyping, the genetic test ordered by Gitti's doctor, enters the picture.

A karyotype offers a snapshot of that fleeting moment when duplicated chromosomes line up in preparation for their incorporation into one of the emerging daughter cells. In a karyotype, these chromosomes can be examined visually, making it easy to determine if any chromosomes are missing or altered.

How is karyotyping performed? The cells used for this type of chromosome analysis are collected from various sources, such as blood, bone marrow, amniotic fluid, or the placenta. They are then placed in a laboratory dish. To visualize the karyotype, these cells are stained with special dyes that highlight the distinct characteristics of each chromosome, creating a unique barcode. This allows for the comprehensive and efficient chromosome evaluation needed to diagnose and treat various genetic disorders.

The first task in the evaluation is to organize the barcoded chromosomes by size. This can reveal not only how many total chromosomes are present but also whether these chromosomes contain all the genetic information they should and whether the material is organized properly. If the patient is revealed to have abnormally organized chromosomes, questions arise. Is the chromosomal arrangement associated with a developmental syndrome? If so, what syndrome is it and how can it be treated?

Two months later, Gitti and her mother came back to my office, and I explained the results of the karyotype.

"This is not trisomy 18," I said. "Gitti's profile reveals something else: there is a very large deletion on one of the two copies of chromosome 18, on its long arm."

I explained how in trisomies, there are three copies of a chromosome rather than the normal two, with the most common—trisomy of the 21st chromosome—causing Down syndrome. The disturbances caused by trisomies derive from excess genetic material, which increases the production of those proteins encoded by the genes present on the extra chromosome. This abnormal increase of some proteins, and not others, disrupts the delicate balance that allows the cell's complex machinery to function as it should.

But chromosomal imbalance can go both ways. Cellular health and function can also suffer when missing chromosomal material leads to insufficient production of certain proteins. Deletion of the long arm of chromosome 18, which contains a large number of genes, triggers wide-ranging developmental consequences. This is because proteins will be generated by only one set of genes, the one inherited from the mother or the one inherited from the father, rather than by both.

Children born with chromosome 18 deficiencies are short in stature and have hearing impairment and other congenital defects. They also generally have some degree of intellectual disability. This is because brain development depends on many molecular processes that rely on the quantities in which certain proteins are produced.

I asked Mrs. Strauss how many of the symptoms associated with chromosome 18 deficiencies seemed to be true for Gitti.

"Almost all of them," she replied. Then I saw her hesitate. Something else was clearly on her mind.

In my years as a geneticist, I have been privileged to get to know patients from many different communities, some of whom have beliefs and values very different from my own. Knowing that Mrs. Strauss and her daughter were members of a Jewish ultraorthodox community in which the ability to bear children is of paramount importance, I thought I might know what was troubling her.

"Are you wondering if Gitti will be able to have children?" I asked gently.

Mrs. Strauss lowered her eyes, clasped her hands, and subtly nodded her head.

"Well, you will be happy to know that there is documentation that yes, some girls with chromosome 18 deletions are able to bear children."

Mrs. Strauss looked up with a spark of hope in her eyes. "Thank you, Doctor," she said.

"This doesn't mean that Gitti will be able to do so. This will need to be addressed when the time comes," I said.

I decided to leave aside the question of which children Gitti might choose to have, understanding that, for Mrs. Strauss, a discussion of choosing healthy embryos that lack Gitti's 18th chromosome deletion during an IVF procedure, something that I would tend to recommend, would be a bridge too far.

Gitti still visits the clinic from time to time. She does need help, but she attends school and is able to do many of her daily chores without assistance. She is obviously progressing, though at a slower pace than others. The medical literature indicates that mothers who have deletions in their 18th chromosome—all of whom have some level of intellectual disability—have a 50% chance of bearing completely normal children, based on their one intact chromosome.

Marriage and childbearing under these circumstances create what many would consider to be an ethical dilemma. Personally, I believe that children and young adolescents with chromosome 18 deficiencies should not be exposed to questions concerning their future parenting and the issue should only be raised if and when it becomes relevant.

Navigating these dilemmas, and communicating them to my patients, is an important part of my clinical work. As a physician and a geneticist, I have the privilege of educating patients about processes that started long before birth but—in every culture and community—can reverberate over an entire lifetime. If my patients leave the office knowing that genetic differences are no one's fault, and with clearer expectations about the future, I've done my job.

Clinical challenges like these, rooted in the fundamental science of molecular genetics, became the topic of many conversations shared by Ehud and Benny over the years. Let's listen to one of these discussions, which took place over two steaming cups of bad hospital coffee.

Ehud: Benny, as a clinician, I know that chromosomal deletions are devastating. From the viewpoint of genetic science, how exactly are these rare deletions created?

Benny: This is a fascinating process. You know that when DNA is replicating, the genetic material is condensed into special structures, with the original material and its copy attached at the middle. This allows the genes to be distributed between two daughter cells, both cells receiving complete and identical DNA content. But while this is what happens in most of our body's cells, it is not the case for the cells involved in fertilization: the sperm for the male and egg in the female.

Ehud: What makes the sperm and egg so fundamentally different in the way they transfer their genes to the next generation?

Benny: The function of these cells is exactly opposite to all other cells in the body, the *somatic cells* that are not involved in fertilization and work hard to pass identical genetic content from one generation to the next. The sperm and egg, by contrast, are designed to shuffle their genetic material, making every newly formed egg or sperm different from the previous generation.

Ehud: But why shuffle? If a certain set of genes carried by eggs and sperm produce healthy offspring, why complicate things? Isn't this shuffling what puts our children at risk for genetic disease?

Benny: You've hit upon the central question of evolution. During sexual reproduction, gene shuffling leads to new genetic variations. This drives the survival of the fittest, the constant selection for the gene-driven characteristics most suitable for survival in a given environment. Just look at man's best friend; there are some 200 different dog breeds, reflecting the traits desired and selected by the people who domesticated them. All these dogs can generate fertile progeny, so whatever advantageous trait emerged because of genetic shuffling is passed down to the next generation.

Ehud: Fair enough. But let's ask the same question in molecular terms. How is this variation generated during the formation of sperm and egg?

Benny: OK, let's trace the process. You know that when a body cell replicates its DNA in advance of cell division, the original chromosomes and the copied versions line up and are later pinched apart when the cell divides. In this process, every daughter cell acquires the same genetic information.

However, during the formation of sperm and egg, a specialized division process generates variability in the genetic material that is acquired by the progeny. For eggs and sperm, the pre-replication lineup looks different: the chromosomes from dad and mom align with each other, but each daughter cell receives only one chromosome of every pair. The deck of chromosomes is shuffled differently in each egg and sperm, and a different combination of chromosomes can be passed down to each daughter cell.

Variability in the composition of daughter cells is further enhanced by an additional molecular mechanism that scientists call *recombination*, which shuffles the genetic material from within a given chromosome. When the chromosomes from dad and mom align, the chromosomes exchange DNA fragments between each other. This shuffling of the genetic deck promotes genetic diversity, which, in evolutionary terms, is helpful for species survival.

Ehud: So, how does shuffling explain chromosomal syndromes like trisomy?

Benny: Sometimes during the recombination phase, one chromosome is shortchanged and receives less DNA than it should. Luckily, this kind of mistake is relatively rare. But when it happens, all the cells of the resulting embryo will carry a chromosome that is missing some of its genetic information.

Ehud: A fateful moment, eh?

Benny: Absolutely. It's something we can't see or control. But it has lifelong consequences.

Chapter 9
Bringing Diagnosis into Focus: The Problem of Missing and Extra Genes

One of my colleagues at the hospital, Dr. Levine, is an excellent pediatrician. I've known her since she was an intern, and even then, despite relatively limited clinical experience, she had terrific instincts, which often led to quick and accurate diagnoses of complex conditions. Recently, however, she called me in to consult on a case that left her at a loss.

"I know this description wouldn't make it into the Physicians' Desk Reference, but this patient," she said, "he just looks funny. He acts funny. I know what I'm saying is vague, but I see no obvious indicators of a genetic syndrome. Can you stop by? Maybe you'll notice something I don't."

Always up for a challenge—especially on a slow day at the clinic—I came over right away. The moment I entered the room, I locked eyes with Billy, a three-year-old boy who broke into a huge grin. But he didn't stop there. Even though the child had never seen me before, he walked over and hugged me around the knees as if reuniting with a long-lost friend. He also started babbling, demonstrating that he was highly verbal but in a manner that, even for someone his age, was somehow off.

"Oh, wow, the red dress is so, so pretty, isn't it? I love red dresses. Red is a happy color! Do you like red? I think red dresses are like apples—so shiny!," he exclaimed, as the opening gambit of a string of non sequiturs.

I crouched down to get a better look at Billy, who continued to smile broadly while he kept up this overly friendly, one-sided, and nonsensical conversation. Billy had low-set ears, and his eyes slanted downward. It was exactly like Dr. Levine had told me over the phone: although Billy was an endearing, delightful child, he looked funny, and he acted funny.

Something about Billy's collection of behavioral characteristics, along with his physical abnormalities, reminded me of a genetic disorder I'd read about but had never personally encountered. Before venturing a premature diagnosis, however, I turned to Billy's mother, who had been observing the child's antics with a benevolent smile.

"Oh, he's a friendly one, all right," she said. Introducing herself as Julie, she launched into a rundown of Billy's history.

"I've been taking Billy to different therapies since about the time he turned two," she said. "You name it, physio, speech, OT, art therapy … he's been there. Luckily, I teach elementary school and my hours are flexible, so I've been able to really devote myself to him."

"Have you ever been referred to a geneticist?" I asked.

"No," Julie answered. "For a while, I've believed that Billy would just grow out of whatever issues he has—he's still so young!—but his behavior raised red flags with the

E. Banne, B.-Z. Shilo, *Dialogue With Our Genes*,
https://doi.org/10.1007/978-981-95-9571-6_9

professionals. If there's something we need to know to make sure Billy gets the best care, I want to know sooner rather than later."

Some genetic conditions have distinct symptoms that make clinical diagnosis relatively straightforward. Others are more mysterious. Similarly, lab tests used to confirm a physician's tentative diagnosis have differing levels of complexity, depending on how the suspected condition would be reflected in the patient's genetic material. For example, the extra 21st chromosome of a child with Down syndrome, or that deleted section of the 18th chromosome in the case of Gitti, are both plain to see in a karyotype test, in which a patient's chromosomes are lined up for visual analysis.

But karyotyping has its limitations. At this resolution, only changes in overall chromosome numbers, or gross chromosomal abnormalities, can be revealed. Further, checking the shape and completeness of chromosomes is a laborious process, requiring finely honed human expertise. In fact, a reasonably competent technician generally needs at least a full year of training to be able to recognize how aberrant chromosomes differ from the normal, healthy arrangement. Since this training is slow, and because karyotyping is expensive, scientists needed a breakthrough—a new method that would make genetic analysis more efficient.

In 1992, at the University of California, San Francisco, Anne and Olli-Pekka Kallioniemi, Daniel Pinkel, Joe Gray, and their colleagues unveiled a new way to read the genome's architecture. Their method—Comparative Genomic Hybridization (CGH)—compared DNA samples to reveal missing or duplicated chromosomal segments, patterns invisible under the microscope. Though created to study cancer, CGH soon transformed clinical genetics, becoming a cornerstone for detecting chromosomal imbalances in patients, especially after the emergence of array-CGH in the early 2000s. The goal of CGH is to identify DNA duplication or deletion, those places in the genome in which the number of copies of a gene is increased or decreased without a proportional change in other genes. The technique is based on a comparison of the quantity of DNA from two subjects, one normal and the other suspected of having a genetic disorder.

In CGH, genetic material is obtained from a healthy person and from the patient needing diagnosis. The genetic material is then tagged using fluorescent chemical markers. The genes taken from the healthy patient are marked in red; the genes of the patient suspected of having a particular genetic syndrome are tagged in green. This procedure makes the genes fluoresce (glow) when they bind to other DNA fragments. The comparison of the fluorescence intensity of the red and green markers reveals any DNA amplification or deletion, which can indicate the presence of a genetic disorder.

Let's pause for a minute. DNA is an enormously complex molecule. How do genetic technicians know where to place their fluorescent tags? The answer to this question emerged from the science of molecular genetics.

As we have learned, DNA is a double-stranded molecule that looks like a twisted ladder. The rungs of the ladder are formed by the connection between two nucleotides (also referred to as bases). Located on either side of the double helix, nucleotides form chemical bonds with other nucleotides in the opposite strand, creating a base pair. But nucleotides don't attach randomly; they are very specific about the partners to which they bind.

As we have seen, there are exactly four types of nucleotides—adenine (A), cytosine (C), guanine (G), and thymine (T)—and every one of these has a dedicated partner. In base pairs, A binds only to T, and C binds only to G. The binding together of these complementary nucleotides is called *hybridization*, and it is the structural glue that holds the two strands of the DNA molecule together. The ironclad rules about which nucleotides bind to each other also create the basis of a diagnostic probe for revealing abnormal hybridization.

In a CGH experiment, scientists use a *DNA microarray*, plates made up of thousands of individual wells, each one containing a single gene. Tagged DNA fragments derived from both subjects are introduced, together, into each well. The tagged genetic material is then mixed and hybridized, meaning the two complementary strands of DNA bind to each other, forming a double helix and also producing a fluorescent signal. The color of this signal reveals which DNA fragment is responsible for the chemical bond, that of the patient or that of the healthy control. The fluorescent signal also allows us to see which regions of DNA are present in what sample and in what quantity.

Another method, called *chromosomal microarray* (*CMA*), eliminates the need for a reference DNA. It uses a special scanner to compare the illumination from the fluorescent signal to the expected illumination. The result is the same. CMA refined the approach even further, detecting subtle chromosomal changes that CGH could not see and doing so without the need for a reference sample.

Both CGH and CMA testing provide important clues for tentatively identifying genetic syndromes. As revealed by the fluorescent signal, higher- or lower-than-normal level of hybridization indicates that the patient has extra copies of genes, or too few. This can lead to abnormal levels of protein production that can cause genetic disease.

> This off-balance fluorescent signal is exactly what was revealed in the case of Billy, my overly friendly young patient. His CMA revealed missing DNA sequences from the 7th chromosome, indicating that Billy may only have one copy of these areas rather than two. It also identified where and on what specific location the deficiency was found.
>
> I called Billy's parents in a couple weeks later. After welcoming Julie back and introducing myself to Billy's father, Ted, I asked them to have a seat.
>
> "Billy's test results showed a loss of genetic material on chromosome 7," I began. "This deletion was found to include 28 genes, some of which code for proteins that carry out essential functions. In Billy's case, these proteins are produced at half the normal level. This explains Billy's condition."
>
> There was silence in the room as my words started to sink in. I continued, "This specific deletion causes a genetic disorder called Williams syndrome."

Before hearing Ted and Julie's reaction to Billy's diagnosis, let's take a moment to define Williams syndrome and see how it relates to the general phenomenon of missing or additional genetic material.

Williams syndrome is associated with a chromosome 7 deletion. One of the missing genes is ELN, which encodes a protein called *tropoelastin*. Tropoelastin proteins bind together to form a larger molecule called *elastin*. Elastin, active in the connective tissue between cells, helps keep fibers stretchy. It is a major contributor to tissues and organs—like joints—that require elasticity to perform their normal function, day after day, and year after year.

Elastin is also an important component of blood vessels, the heart, and skin. A genetic deletion leading to a shortage of tropoelastin causes a downstream reduction in elastin. Skin with only half of its normal elastin content has a reduced ability to recoil after stretching and will be lax and soft to the touch. A total loss of elastin, as seen in the very rare Cutis Laxa syndrome, causes skin to be extremely lax and stretchable. This shortage gives rise to many of the syndrome's hallmark features—from soft, pliable skin to the vascular stiffness that can lead to heart problems. For patients with Williams syndrome, the situation is less dire; they have half of their elastin intact.

My use of the word *syndrome* clearly shook both parents. While Ted remained silent, it triggered a flood of questions—or, more accurately, a protest—from Julie.

"Are you sure?" she asked. "Is it like Down syndrome? In so many ways, Billy is completely normal! Ted, tell the doctor what Billy's teachers have been saying! He's been making such progress!"

There was no point in answering immediately, so I waited until the atmosphere in the room, while still tense, calmed down enough for me to address Julie's concerns.

"Billy has a collection of symptoms that, when put together, are called a syndrome. This is not Down syndrome. It's just a way to refer to Billy's particular set of differences and give it a name."

I then started asking them about other symptoms characteristic of Williams syndrome. Did he have a heart condition? Yes, an atrial septal defect that had been repaired by catheterization. He was currently being monitored for aortic stenosis, a cardiovascular disease. Billy had two additional symptoms on the Williams syndrome checklist: a recent diagnosis of poor eyesight requiring glasses and a kidney abnormality discovered during an evaluation for renal disease.

"It seems to me that you've been very conscientious about having your son seen by specialists and cared for, and that's certainly one of the reasons Billy is as healthy and happy as he obviously is," I told them. "You seem to have done everything right. All that was missing was the genetic diagnosis."

Ted spoke up for the first time, and I could see tears well up in his eye. "Julie gets the credit for that," he said. "With the hours I work, I haven't had much time to be involved."

Julie reached out to squeeze Ted's hand, then turned back to me. "What does this mean for Billy's future?" she asked.

We discussed the clinical characteristics of Williams syndrome for a while, including what Billy had and what he didn't have, as well as what needed follow-up. Cognitive impairment is universal in Williams syndrome, accompanied by a lack of fear of strangers, overfriendliness, and the challenges individuals like Billy face in achieving independence, which most do not reach. We talked about the anxieties that Billy already experienced, the high risk of hearing loss, and his short stature. Our conversation lasted for a long time.

Addressing both parents, I began what some might think was a futile line of conversation, relating to the possibility that Billy might eventually be able to father children. I felt that even if this is a remote option, the parents have the right to know the potential consequences.

"If and when Billy marries, his children will have a 50 percent chance of transferring this deletion to his children," I said.

Watching an additional measure of sadness wash over their faces, I told them how such a result could potentially be controlled through pregestational testing for the deletion, by examining an early-stage embryo fertilized through IVF.

So where does this leave us?

When scientists and physicians examine genetic material for defects, they start with the most general view—the chromosomes as laid out in a karyotype. Based on the karyotype, they may progress to more refined tests, such as CGH or CMA, to identify excess or missing genetic material. Further research, using additional testing methods, can reveal a patient's genetic profile in more detail.

It can be helpful to think of this genetic sleuthing in terms of something that held a place of pride on the shelves of every pre-Internet library and in many homes: the physical volumes of an encyclopedia.

The huge amount of genetic information packed into the nucleus of each of our cells can be likened to 1000 encyclopedia volumes. When we examine chromosomes, this is like tallying up the volumes to see if one is missing. If all the chromosomes are present and accounted for, we move to the next level of detail. We count the number of pages in each volume, to see if there are missing pages.

In Billy's case, all the chromosome volumes were present and accounted for, but CGH testing revealed missing pages, in the form of 28 missing genes within chromosome 7.

Other genetic technologies make it possible to dive deeper still and to find differences and defects that are more subtle. Scientists can identify changes in a single word on a page, or even a single letter, within the vast 1000 encyclopedia volumes of our genetic inheritance. How is this possible? Read on to find out.

Chapter 10
The Diagnostic Deep Dive: The New Genetic Technologies Providing Answers

It's a clinical test that seems diabolically designed to drive anxious parents over the edge: holding the newborn carefully, the examining pediatrician executes a sudden arm movement that, for a split second, allows the baby to drop down a bit before being securely caught again. Most infants will respond by flinging their arms out to the side. Some will start to cry. But despite the momentary discomfort, this startle—also known as the *Moro reflex*—is good news. It demonstrates that the baby's brain and nervous system are developing normally.

The startle reflex generally disappears by 2 months of age and is forgotten by the parents, who can then turn their attention to the thousands of other worries associated with childrearing. But as a visitor to my clinic described it, her infant son's startle reflex remained a focus of medical attention for quite a while.

> "Mark's startle reflex was more dramatic than expected, and our pediatrician said that this was worth checking out because his behavior can sometimes hint at a genetic syndrome," Anna recalled, sitting in my office with her husband Dima and, together, keeping an eye on their son playing on the floor. "She recommended having Mark tested for Tay-Sachs. Luckily, the test came out negative."
>
> *Infantile Tay-Sachs* is a devastating genetic disorder that results in the destruction of nerve cells in the brain and spinal cord. Although very rare in the general population, the mutations that cause this disease are most common in people of Ashkenazi (Eastern and Central European) Jewish heritage like Anna and Dima, whose Jewish parents had immigrated to Israel from the former Soviet Union. But while children diagnosed with Tay-Sachs develop motor disabilities in early infancy and generally die before the age of four, Anna and Dima's son was 5 years old.
>
> It's worth noting that the suspicious startle reflex Anna and Dima described didn't sound nearly as intense as what we observe in children with Tay-Sachs. I also didn't see any evidence of this type of behavior during the family's visit to the clinic. At the same time, it was clear that Mark, who was still wearing a diaper and had linguistic capabilities characteristic of someone much younger, had systemic developmental delays.
>
> Dima took up the story. "The startle reflex was just part of the picture," he said. "From the time he was born, Mark had tremors. He was tested for epilepsy, but the test came back negative. Our doctors—and we've seen many of them over the years—decided to treat him

E. Banne, B.-Z. Shilo, *Dialogue With Our Genes*,
https://doi.org/10.1007/978-981-95-9571-6_10

with antiseizure medication. It didn't help. His symptoms got worse. The tremors could be scary, causing his eyes to roll back in his head."

"That's why we were referred to you," Anna continued. "Mark has a lot of different symptoms, but so far, no one has been able to put their finger on what causes them."

When a patient displays physical abnormalities, the path to a genetic diagnosis can sometimes be relatively straightforward. Mark presented a different type of challenge.

The first step was to send Mark for chromosomal microarray (CMA) testing. CMA can identify duplicated or missing chromosomal segments like the trisomy, or extra copy of chromosome 21, seen in Down syndrome. However, in Mark's case, the CMA testing came back normal. We needed to dig deeper.

For investigations that need higher levels of detail, geneticists use an approach called *next-generation sequencing*, or *NGS*. Based on scientific research performed at the University of Cambridge—incidentally, in the same British city where Watson and Crick characterized DNA's double helix structure—NGS makes it possible to read extremely large volumes of DNA sequences.

To review, a DNA sequence is the order in which the nucleotides adenine (A), cytosine (C), guanine (G), or thymine (T) appear in a DNA molecule or fragment. As opposed to microarray methods, which can reveal chromosomal abnormalities, NGS makes it possible to identify variations in single nucleotides.

NGS technology is relatively new, dating back to the very first years of the twenty-first century. At the time, scientists and bioengineers were pondering the next steps beyond the Sanger sequencing used in the Human Genome Project. Although the data generated by Sanger sequencing was impressive—it could be used to analyze 96 strands of DNA in parallel—this approach was inefficient and very expensive. Based in part on its reliance on Sanger sequencing, the Human Genome Project took 13 years to complete and cost a staggering $3 billion.

In the waning years of the project, Nick McCooke, an engineer with background in business, had an idea. He was leading a team at a company called Solexa, which was exploring the applicative potential of the basic science breakthroughs that the Human Genome Project had produced. Together with his coworkers, McCooke developed a new instrument that could simultaneously analyze tens of millions of genomic fragments. This advance turbocharged sequencing speed while bringing costs way down.

How did they do it? As opposed to Sanger sequencing, in which only a single DNA strand is analyzed at a time, in McCooke's method, DNA is broken into fragments made up of just tens of bases (remember, *nucleotides* and *bases* are two words for the same thing). Thousands of these fragments are stuck randomly on a single glass slide, where they are amplified; each fragment is duplicated, so that it appears on the slide in many identical copies. Then, in a lightning fast robotic process, fluorescent bases are added to the slide, one at a time. The fluorescent bases act as probes; when they attach to a DNA fragment, a glowing signal is generated, revealing their exact binding location.

Because bases bind selectively (A binds only to T, and G binds only to C), McCooke team's new technology revealed the exact order of bases in each fragment. The glowing signals were recorded, and then the fluorescent labels were chemically removed so that the cycle could be repeated.

McCooke's new method based on short DNA fragments generated more nucleotide-level genomic information than had ever been achieved.

Today, *next-generation sequencing* is a catchall term for several specialized analytic technologies that are helping to advance precision diagnostics and personalized medicine. To pinpoint the genetic story behind Mark's unique collection of symptoms, NGS was the obvious next step.

When I told Anna and Dima that their son's microarray results were inconclusive, they received the news stoically. Apparently, their experience taking Mark to so many doctors had taught them that there would be no quick fix and that patience was a virtue.

For the next step in our diagnostic investigation, I suggested a next-generation technique called *whole exome sequencing*.

"Whole exome sequencing is a lab method that can identify very subtle changes in the genetic code," I told them. "It focuses only on 1 percent of our total DNA, which is called the *exome*. It's the part of our DNA that carries the genetic code for creating functional proteins."

"In other words, we'll be searching for a needle in a much smaller haystack," Dima remarked.

I smiled and then described the protocol. "Actually, we'll be sifting through three haystacks," I said. "In whole exome sequencing, we compare the child's genome to the genome of both parents. It's an efficient way to get tens of thousands of results and identify those gene variations that warrant further investigation."

Anna and Dima exchanged bemused glances. "OK," Anna said with a little sigh. "Let's get started. How do we make the appointment?"

This positive attitude would help the young couple persevere through yet another setback. Two months later, we received Mark's whole exome sequencing results. The test indicated that Mark's genetic profile was, on the nucleotide level, pretty much what one would expect based on his parents' DNA. Nothing abnormal was found.

"Exome sequencing is powerful, but it may miss nucleotide variations in sections of the genome that have not yet been identified as containing protein-producing genes," I said. "That means that some genetic regions have probably been lumped in with what was called "junk DNA"—that does not code for proteins. Also, exome sequencing may not be ideal for finding differences in genome structure."

Aiming for encouragement, I added a positive note. "There is some good news, though: the test you've done has generated a great deal of data that can be interpreted using other tools. That's what we'll try next."

Mark's case was an example of those situations where my work as a physician and clinical geneticist becomes most interesting. Because the field of genetics is continuously evolving, I cannot assume that the answer to my patients' questions will be found in any standard textbook or even in recent scientific publications. Sometimes, it becomes necessary to advance the science on my own.

At the time, I worked in a hospital associated with the Hebrew University of Jerusalem. One of the advantages of working in a university hospital is the opportunity it affords to bring clinical problems to scientists working in university research labs. With the help of these scientific colleagues, I used two online tools designed to sift through and interpret whole exome testing data. Our goal was to identify how

Mark's nucleotide sequence varied from that of his parents and to see whether such variations might have functional impact.

The online tools we employed were based on an artificial intelligence (AI) algorithm. But the human aspect was still important. When the AI software highlighted a variant that might potentially be the source of the child's symptoms, I took this as the jumping off point for further research.

The software revealed the mutation that eventually led to our eureka moment: Mark had a novel variant in one of his two copies of a gene called *STXBP1*. This variation would have been easy to miss. Just one amino acid (a molecule that is one of the building blocks of proteins) was duplicated.

During my years as a physician and geneticist, I'd learned that mutations to the *STXBP1* gene are associated with a very severe form of epilepsy. But in Mark's case, epilepsy had already been ruled out. Could an extra amino acid in one of Mark's *STXBP1* genes cause something else? Digging into the literature, I found that such a mutation might interfere with the gene's normal function—a function essential for producing the proteins that allow brain cells to communicate with one another.

STXBP1 produces a protein that mediates signal transduction, the process by which chemical messages are transmitted through the gap (the synapse) that separates one neuron from another. The process is mechanically fascinating. The first neuron packs the message into a little protective bubble called a *vesicle*. Once the vesicle arrives at the synapse, it merges with the second neuron's membrane, a mechanical process known as *docking*.

One of the most important players in this docking system is—you've guessed it—the *STXBP1* gene. The protein this gene produces makes it an essential component of the nervous system's communication network.

It's not clear exactly what functional abnormality was created by the extra amino acid in Mark's *STXBP1* gene. However, there was no question that any disruption of this fundamental neural communication system could have a catastrophic effect. This made our discovery important not just for Mark and his family but also for other children whose abnormalities might go undiagnosed in the future. In 2020, my colleagues and I published our findings in a scientific journal.

Not only did the sophisticated NGS tools we used help us identify the involvement of *STXBP1* in Mark's condition, but they were also very important to our research progress. In addition, intuition based largely on our clinical experience played an important role. The *STXBP1* abnormality provided circumstantial evidence of a genetic cause.

Back in the clinic, I summed up the findings and Dima asked the obvious question.

"If this mutation causes Mark's condition, where did it come from? Earlier, you said that our genes—Anna's and mine—were normal. Could this still be something we gave him?"

Recognizing the concern in Dima's voice, I reassured him that Mark's condition could not have been foreseen or prevented.

"Sometimes a genetic alteration arises for the first time during formation of the egg or sperm of one of the parents," I said. "This is called a *de novo mutation*, based on the Latin word for new. Frankly, it's surprising it doesn't happen more often."

"Why surprising?" Anna asked.

"Well, the process of creating eggs and sperm, what scientists call *germ cells*, is enormously complicated," I said. "In females, the number of cell divisions to generate eggs is limited. But in the male, continuous cell division generates a fresh supply of sperm throughout men's lives. Some 1,500 sperm cells are generated each second, and the process is prone to mistakes. If one of the precursor cells carries a mutation, the mutation can be passed on during fertilization. As opposed to mutations that can be detected in every cell of the body, this kind of variant, called a *germline mutation*, cannot be identified by routine blood testing of the parents. This is because it is carried by only a small fraction of cells that produce the sperm or the eggs."

"So will this germline mutation affect children we might have in the future?" Dima asked. "And what does this mean for Mark?"

"These are excellent questions," I responded. "The bottom line is that when you decide to have additional children, prenatal testing will be important."

I hastened to add that since Mark's mutation was a rare random event, the chance of having another child with the same syndrome was very slim. "As for Mark, the good news is that the science of clinical genetics is advancing very quickly. There are new discoveries every day. I can't predict if a new drug will emerge that could be used to treat Mark's condition. But I would like you to come back for an annual checkup. I'll keep track of Mark's development and keep you informed of any new therapies that may emerge."

Chapter 11
From "Bruiser" to Prospective Dad

As a doctor, I always try to go the distance. Maybe that's why I'm inspired by the story of James Lind, a British surgeon who served aboard the *HMS Salisbury* in the mid-eighteenth century. While on patrol in the North Atlantic, Lind performed one of the medical world's first controlled clinical trials, treating sailors suffering from scurvy.

Scurvy is a life-threatening disease that causes swollen gums, uncontrolled bleeding, and easy bruising. Trying various remedies, Lind found that sailors who ate citrus fruits soon recovered. His published findings led the British Navy to add lemons to the on-board menu (later replaced by cheaper limes, source of the derogative nickname for British sailors, "limeys").

Many years later, laboratory researchers revealed the science behind this maritime success story, determining that scurvy is caused by vitamin C deficiency. Lack of vitamin C affects the production of collagen, the most abundant type of protein in our body, comprising about 30 percent of all our proteins. Collagen provides structural stiffness to many different body tissues. If collagen is not continually maintained and replaced, tissues break down. That's what happens in some collagen-related syndromes that I occasionally see in my clinical genetic practice.

As a pediatrician, most of my patients are children. But 1 day, I met Jeremy, an adult patient suffering from a collagen-related disorder, together with his wife Maria.

> No stranger to genetic testing, Jeremy knew that it was his genes—not his diet—that caused his symptoms. As the couple entered the room, I suspected the same, immediately noticing Jeremy's loose-jointed gait, sunken eyes, and smooth, wrinkle-free forehead. He also had very slender fingers that seemed mismatched with his hefty torso.
>
> Given Jeremy's constellation of physical symptoms, I was tempted to order a set of tests right away. But as my long years of clinical practice have taught me, careful listening is the foundation of any successful consultation, as well as the basis for building trust. So I invited Jeremy to tell his story.
>
> "For as long as I can remember, I've been active in sports, and my dad used to call me his tough little bruiser," Jeremy began. "But the fact is, I bruise really easily, and when I bleed, my wounds don't heal like they should."

E. Banne, B.-Z. Shilo, *Dialogue With Our Genes*,
https://doi.org/10.1007/978-981-95-9571-6_11

Jeremy reported that his father and sister had similar symptoms. He also said that while he was aware that his condition had a genetic basis, it never restricted his sports activities or his professional advancement as a security guard. He also never let it affect his decision-making—until now.

"I enjoy playing football and figure that I'm tough enough to handle any bruise or slow-healing scrape that comes along," he said. "But Marie and I are hoping to have kids." He paused to glance at his wife, then turned back to me. "We know that this condition can be dangerous, and we don't want to pass it down to our children."

Further discussion revealed that Jeremy was born with bilateral clubfoot, a deformity in which a baby's feet are rotated inward and downward. Long after the problem was corrected with splints, Jeremy continued to have balance issues. His family history was also a bit concerning. Several family members had experienced sudden death. He had an aunt who experienced a uterine tear during pregnancy that caused her to lose the baby.

"I'm glad you came to see me today," I said. "The symptoms you describe could indeed have a genetic cause. Therefore, there may be decisions you need to make before you decide to have children. Even though you've been tested before, I'd like to order up some lab work, so we can find out exactly what's going on."

Jeremy's symptoms were typical of a genetic defect that affects the production of collagen. Although he insisted that this did not limit him in any way, I knew that Jeremy's luck might run out at any moment.

Collagens are critical for creating the tissues and structures that allow our body to function. They determine the level of mineralization in particular tissue types—from bone, which is hard and stiff, to tendon, which is soft and pliable, to cartilage, which is something in between.

Given that collagens make up a substantial portion of our body's proteins, any defects in collagen production can have severe consequences. The heart relies on collagen for various functions such as providing resistance during heart expansion, insulating electrical signals that travel between heart chambers, and determining the physical characteristics of cardiac valves.

People with collagen abnormalities, therefore, have more to worry about than bruises and bleeding. Some of these conditions can lead to the sudden rupture of vital internal organs. This poses a chronic health threat that can have fatal consequences.

Jeremy had something in common with those sailors from long ago who suffered from scurvy: his life situation. His was a life of constant, imminent, and life-threatening danger, even if he was not fully aware of it. It also reminded me of another historical story, this one passed down by Cicero, a Roman statesman and philosopher from the first century BCE.

Cicero retells an ancient parable concerning Dionysius II, the tyrannical ruler of the city of Syracuse. Dionysius wanted to torment Damocles, a member of the royal court who, out of envy for his master's wealth, offered to take the king's place for a day. Surprisingly, Dionysius agreed, placing Damocles on his own golden couch and treating him to succulent cuts of meat and other delicacies.

Damocles was thrilled with his good fortune until he looked up and noticed that the king had suspended something from the ceiling: a razor-sharp sword. Positioned directly over Damocles's head, the sword hung by a single strand of horsehair, which threatened to give way and kill him at any moment. Thus, the *sword of*

Damocles entered into the lexicon as shorthand for what my patient Jeremy was experiencing without being fully aware of it: a life lived in the shadow of imminent death.

Cicero's writings are remembered for something optimistic as well. In a letter to his friend Atticus, Cicero famously stated, "While there's life, there's hope."

> Jeremy and Maria wanted to be sure they would give birth to a healthy baby. This is every prospective parent's wish, of course, but Jeremy's genetics complicated matters. Another complication was bureaucratic: the results of genetic tests performed on Jeremy when he was a child were no longer available. I told the couple that there was actually an advantage to starting the process from scratch.
>
> "Back in the early days of genetic testing, not so long ago—the 1980s or 1990s—the pathway to diagnosis wasn't as easy as it is today," I told them. "Doctors would begin with a working diagnosis based on an educated guess about what gene or genes might be causing the problem. They would then look for mutations in just these specific genes, one at a time, even though the real source of the problem might be elsewhere."
>
> "Like searching for something you've lost only under the streetlight where it's easier to see," Marie remarked.
>
> "Exactly," I said. "Nowadays, genetic techniques allow us to screen the entire genome without prior assumptions, looking for alterations in single nucleotides, the bases C, A, G, and T that are the building blocks of inheritance. If we find something surprising in the sequence—say, there's an *A* for the adenine, where we would expect a *C* for cytosine—that's a start. We can use this clue to compare different members of a single family. If the deviation appears in family members who have similar symptoms, like Jeremy, his father, and his sister, yet does not appear in healthy family members, the mystery is solved. We'll have identified the single nucleotide alteration that causes the syndrome. This type of analysis will be essential for you as you plan your pregnancy."
>
> "How so?" Jeremy asked.
>
> "After we've identified this unexpected change—what scientists refer to as *a single nucleotide change*—it becomes possible to take action and prevent its transmission to the next generation even without fully understanding the underlying molecular cause."
>
> "Take action," Maria repeated the phrase slowly to herself, then looked at me. "Take action ... when?"
>
> Recognizing the world of anxiety behind Maria's question, I answered quietly. "With careful family planning," I said, "it is possible to prevent the birth of a child who carries Jeremy's genetic syndrome, even before you become pregnant."

Single nucleotide polymorphisms, referred to by the acronym *SNP*s and pronounced "snips," are the most common type of genetic variation. In every SNP, one nucleotide—either adenine (A), thymine (T), guanine (G), or cytosine (C)—is replaced by a different one that does not usually appear in that position.

Scientists have found more than 100 million SNPs in humans, and most of them have no effect on health or development. But certain SNPs can be used to track the inheritance of disease-related genes within families.

Scientists have identified SNP-related patterns that serve as markers for specific diseases. Today, all these markers are incorporated into a single laboratory test that analyzes the genome, revealing which disease-related SNPs are part of a person's genetic inheritance. When the results of this test are compared to those of other family members, the presence of the same SNP pattern in those with the disease and its absence in healthy family members provides strong evidence for the SNP(s) being the cause of the syndrome.

Why does such a small genetic change have such a major health impact? Not all changes in DNA are the same. Scientists use the term single nucleotide polymorphisms, or SNPs, to describe common variations—small differences in a single "letter" of DNA that most often have no effect on health. But sometimes, a change occurs that is not common and does have consequences. To keep things simple, we will refer to such a change as a single nucleotide change, or SNC—a rare alteration in a single DNA letter that can disrupt the function of a gene. Like any meaningful alteration to our DNA, an SNC can affect how a gene is translated—that is, how it gives rise to the proteins that carry out biological functions. In some cases, the resulting protein may be shortened, unstable, or not produced at all.

And if the SNC affects the production of a critical protein like collagen ... well, let's get back to our story.

When Jeremy and Maria next appeared in my office, I had definitive news.

"Your test results came back, and the answer is very clear: you have a single genetic mutation, a snip that has been passed down to some but not all members of your family. This mutation is the cause of your symptoms. It affects the COL3a1 gene and causes vascular Ehlers-Danlos syndrome, or vEDS. This is a collagen-related disorder."

Although collagens are produced inside cells, they carry out their function—creating structures with the proper stiffness and mechanical features so that organs can function properly—after they are secreted through the cellular membrane into the surrounding environment.

Collagen production involves multiple genes. But if, as happens in vEDS, one of these genes is affected by a single nucleotide polymorphism, the collagen will come out wobbly and the collagen-based structure will not stand.

Collagen proteins are very long and the DNA that encodes for them consists of a long sequence of nucleotides where mutations can occur. Despite molecular safeguards that usually allow collagen production to proceed normally, mutational changes can, and do, occur at critical junctions. Such genetic aberrations lead to wobbly collagen, the same type of malady that those unfortunate sailors suffered from due to a lack of vitamin C.

Not all of my patients are interested in the molecular story behind their particular syndrome. But in this case, Jeremy and Maria were anxious to understand exactly what they might pass down to their future children. So I went on to describe collagen in detail.

"Normal collagen is made of three very long protein chains that twist around each other," I began. "In vEDS, however, due to a mutation in a single nucleotide, only two of these three proteins are produced properly and in the correct amount. The third protein is compromised by the mutation. This reduces the overall strength of the twisted collagen."

I explained that different mutations can lead to abnormal collagen chains. I added that some of these mutations produce symptoms—like Jeremy's tendency to bruise easily—that are not very serious. But in other cases, collagen disorders are a true sword of Damocles, creating a constant and very serious threat to health and even survival. This is the threat that Jeremy had lived with from childhood, without ever being aware of it.

"The danger to people who have vEDS is that collagen structures are not as strong as they need to be," I said, keeping my voice steady and low and watching for Jeremy and Maria's reaction. "This means that major internal organs, like the arteries or intestine, could rupture. This is something that, theoretically, could happen at any time."

There was silence in the room, as the young couple took in this information. Jeremy was the first to recover.

"I've managed to live with this up until now, even though I didn't know its name," he said. "But let's put me aside for a minute. What does this mean about having children?"

"Let's begin by saying that it is extremely lucky that you, and not Maria, have the diagnosis," I replied. "VEDS puts the structural integrity of the uterus—and every other hollow organ in the body—at risk. But there is good news: in terms of pregnancy and childbirth, the situation can be managed. If the two of you are willing to plan your pregnancies carefully, the technology exists to ensure that none of your future children will inherit this genetic mutation."

I gave Jeremy and Maria a referral for a consultation with an expert in pregestational genetic testing, or PGT. Prospective parents who choose PGT commit to having only protected sexual intercourse and to relying on in vitro fertilization to create early-stage embryos. Before a fertilized egg is selected for implantation in the mother's womb, it is screened for the presence of harmful mutations—in Jeremy and Maria's case, the mutation that causes vEDS.

As we wrapped up our session, Maria, looking visibly troubled, took her husband's hand.

"I wish that we could just go ahead and try to get pregnant without any particular restrictions," she said. "Still, it's good to know that there is a way to take control of the situation. My nightmare was getting pregnant, doing amniocentesis, and finding out that our child has Jeremy's syndrome. Then, we'd have to decide whether to abort. Hopefully this is a decision we'll never have to make."

As I handed the couple their referral and wished them good luck, I knew that their case, like so many in the genetic clinic, had layers of unexplored ethical and emotional issues. These were the kind of questions that made my work so fascinating. As a geneticist, I can give advice. But ultimately, each family must find its own answers.

Chapter 12
Genetic Tests During Pregnancy

In the previous chapter, we left Jeremy and Maria with a big dilemma. Jeremy was diagnosed with a heritable genetic disease, and the couple wanted to know if it was possible to avoid transmitting the faulty gene to future offspring and, if so, how.

This question is important to our understanding and management of all heritable genetic diseases. Sometimes, such conditions can be identified in the parents before any children are born. In other cases, parents discover they are carriers only after their child is diagnosed. Once the risk of genetic disease has been identified, parents' desire to have as much information as possible is understandable, given the statistics: about 3% of pregnancies in the United States will culminate in a child with a severe physical malformation and/or mental disability.

In dealing with this challenge, the work of the geneticist or counselor varies depending on the patient. For children carrying symptoms of a condition, a professional genetic workup can help achieve a clear diagnosis. It can offer the parents emotional support while guiding them toward the best practices for their child's optimum health and well-being. The process can also provide vital prenatal information. It can help prevent the birth of babies who have a high likelihood of serious genetic disease, while improving parents' chances of having healthy offspring.

There are ethical issues involved in any decision related to preventing or terminating pregnancy. This is obviously also true for the decision to give birth to children known to be at risk for genetic disease. These complex ethical questions will be dealt with in detail in a dedicated chapter. Here, we would like to focus on something else: the options for parents who seek out prenatal genetic analysis.

Before we dive in, it is important to note that genetic counselors can provide essential support for pregnancy preparation. Even before pregnancy, they can offer prospective parents screening tests to make sure they are not carriers of any known genetic conditions. But even after a couple has gotten the genetic green light and become pregnant, prenatal testing may be required. It can help maximize the chances of carrying a healthy child to term.

E. Banne, B.-Z. Shilo, *Dialogue With Our Genes*,
https://doi.org/10.1007/978-981-95-9571-6_12

Not every prospective parent is familiar with the various genetic tests performed during pregnancy. To clarify the options, the various procedures, their goals, and their health implications are laid out in the form of Frequently Asked Questions, all of which are answered below.

12.1 FAQ: Prenatal Genetic Testing

1. Why perform prenatal genetic analysis?
2. Who should be tested?
3. What prenatal genetic test can be done early in pregnancy?
4. What about prenatal genetic analysis later in pregnancy (amniocentesis)?
5. What about testing the umbilical cord?
6. What are the risks of prenatal genetic analysis?
7. How are genetic tests done during IVF, prior to embryo implantation?
8. Who makes the decision about testing, the parents or the doctor?

12.1.1 Why Perform Prenatal Genetic Analysis?

Genetic tests during pregnancy focus on the fetus, and their goal is to identify potential abnormalities. These tests are described as "invasive," as they involve examining the genetic material of the fetus in utero—that is, in the mother's womb.

Parents decide to authorize such tests because they want answers to critical questions: Is there anything wrong with my fetus? Did my fetus receive a defective gene? What caused the abnormality indicated by previously performed tests? What's the next step?

12.1.2 Who Should Be Tested?

If testing of the parents prior to pregnancy reveals a genetic condition, it is necessary to test the fetus for the specific genes that cause the condition. In addition, it is possible that routine pregnancy monitoring with ultrasound will reveal a possible abnormality that may be associated with a genetic defect. This would prompt a referral for testing.

12.1.3 What Prenatal Genetic Test Can Be Done Early in Pregnancy?

The genetic test that can be done earliest in pregnancy is known as *NIPD*, or *noninvasive prenatal diagnosis*. Available as early as the 9th week of pregnancy depending on the clinic, NIPD is a blood test that screens for birth defects by analyzing fetal DNA in the mother's blood. It can screen for conditions such as Down syndrome and trisomies 13 and 18. It can also determine the baby's sex, some disorders related to sex chromosomes, and some microdeletions. If NIPD testing comes out negative, that means that there is low risk for genetic conditions in the fetus. A positive result may indicate a need for further tests, like ultrasound or amniocentesis. Because it is a simple blood test, NIPD may eventually replace the tests now typically used in the first and second trimesters. NIPD can also screen for certain single-gene disorders, including Alagille syndrome and achondroplasia, particularly in pregnancies with a known genetic risk.

The earliest invasive genetic test is called *chorionic villus sampling*, or *CVS*. In CVS, the placenta, rather than the fetus, is tested. Since the placenta begins as part of the fetal tissue, placental genetics closely mirror the fetal genetic profile. This test is usually performed between the 11th and 14th week of pregnancy, and the timing is critical; undergoing this procedure before the 10th week of pregnancy poses a risk to the fetus's limbs and should be avoided.

In CVS testing, the physician targets the point at which the placenta meets the uterus, called the *chorionic villus*. The physician aspirates the tissue and extracts some of the cells, which go through a separation procedure to isolate the fetal cells. These fetal cells are amplified in cell culture to make sure there are enough for all the molecular tests needed. They include standard chromosomal microarray analysis (CMA) and may also include a specific molecular test focusing on genetic factors under investigation in the fetus and family.

CVS usually provides an accurate representation of fetal DNA and is a strong indicator of the presence of genetic abnormalities such as those that lead to Down syndrome. Because they know the results at a very early stage in pregnancy, parents can decide how they want to proceed. Some will continue the pregnancy and look for clearer expressions of the syndrome. Some parents will choose to terminate.

One of the things that makes this such a challenging decision is the fact that spontaneous variations in the genetic material can sometimes cause differences between placental and fetal genes to occur. CVS, which uses placental cells to identify genetic variations in the fetus, sometimes produces results that do not truly reflect the fetal genotype.

One situation that can lead to this ambiguous result is called *confined placental mosaicism* (*CPM*). In CPM, some cells in the placenta have an abnormal number of chromosomes, while the chromosomes in fetal cells are normal. Identified in about 1–2% of pregnancies during CVS screening, CPM can sometimes result in early pregnancy loss. However, skilled genetic specialists can often determine when chromosomal discrepancies are limited to the placenta and do not impact the developing baby.

12.1.4 What About Prenatal Genetic Analysis Later in Pregnancy (Amniocentesis)?

Amniocentesis, performed between the 17th and the 22nd week, is the gold standard invasive genetic test done during pregnancy. In this procedure, amniotic fluid is aspirated from the sac in which the fetus resides. This fluid contains fetal cells, which may originate from the developing baby's digestive system, skin, kidneys, bladder, and genitals. These cells are examined in much the same way as described above. Amniocentesis can therefore detect almost all chromosomal and genetic disorders, depending on the test performed.

Like CVS, amniocentesis has its limitations. Its findings are based only on the cells that are gathered and used for testing. Therefore, it cannot detect the presence of tissue-specific mosaics, a phenomenon in which cell lines derived from the same fertilized egg, found in different parts of the fetus, have variations in their respective DNA. The geneticist or genetic counselor will discuss these limitations with patients when relevant.

12.1.5 What About Testing the Umbilical Cord?

After around the 18th week of pregnancy, another invasive genetic test becomes possible called *percutaneous umbilical blood sampling* (*PUBS*). Also known as cordocentesis, fetal blood sampling, or umbilical vein sampling, this quick test examines fetal blood directly from the umbilical cord to detect disorders in the fetus.

Like the blood tests performed on adults, umbilical blood sampling appears to be an ideal method for investigating genetics. However, the risk of pregnancy loss varies between 1% and 2%, according to various studies. This risk is nearly 10 times greater than the risk associated with amniocentesis and about 5 times higher than that of CVS. Therefore, PUBS is typically conducted only when other procedures, such as CVS or amniocentesis, are not feasible or when fetal blood cells are needed for diagnostic testing.

12.1.6 What Are the Risks of Prenatal Genetic Analysis?

CVS testing is generally safe but comes with some risks, which expecting parents should discuss with the genetic counselor and the physician. These include the possibility of miscarriage, which occurs in less than 0.5% of patients. As with most invasive medical procedures, there's a small risk of infection. There is also a small risk that the procedure will result in inaccurate test results.

Like CVS, amniocentesis procedures are considered safe. But pregnant women and their loved ones should be aware of potential risks that, in a small minority of cases, can affect the mother and the fetus. Complications of amniocentesis can include cramping, bleeding, and leaking amniotic fluid. In less than 0.2% of cases, amniocentesis leads to miscarriage or early delivery.

12.1.7 How Are Genetic Tests Done During IVF, Prior to Embryo Implantation?

A completely different track of genetic fetal-embryo testing relies on in vitro *fertilization*, or *IVF*. This method, credited to Nobel Prize–winner Sir Robert Geoffrey Edwards, was initially developed to help couples struggling to get pregnant.

In IVF, a woman's egg is fertilized by a man's sperm in a test tube. Because women usually produce only one egg at a time, hormones are used to stimulate the ovaries to produce more eggs. This is to ensure that when the eggs are aspirated from the ovary—under sedation, in an operating room setting—enough eggs will be retrieved to promote success and avoid having to repeat the procedure.

The male part of IVF is simpler. The man is asked to masturbate and collect the sperm produced in a tube. After this, the retrieved egg is placed in another tube, and the extracted sperm is added to fertilize the egg.

In some cases, just mixing the sperm and the egg is not enough. A specialized procedure called *intracytoplasmic sperm injection* (*ICSI*), which is an assisted reproductive technique, is needed. In ICSI, a single sperm is directly injected into an egg to facilitate fertilization using a fine needle under a microscope. If fertilization occurs, the resulting embryo is cultured for a few days before being transferred into the woman's uterus to establish a pregnancy. This method is particularly useful for couples experiencing male infertility issues, such as low sperm count, poor motility, or abnormal sperm morphology.

In IVF, the fertilized egg is kept in culture, where it divides until enough cell divisions produce a stable embryo. Ultimately, the early-stage embryo will be implanted in the woman's uterus in hopes of bringing a healthy baby to term. Under normal circumstances, when a genetic syndrome is not suspected and searched for, the success rate for IVF is approximately 40% per cycle.

But for parents who know that the children they bring into the world are at risk of inheriting a genetic mutation, things are not so simple. If they have already given birth to a child with a genetic abnormality, or if they carry a genetic disorder themselves, they need to manage the risk of embarking on a pregnancy that can result in a child born with their genetic disease.

In such cases, preimplantation genetic testing can allow parents to choose an embryo that does not carry the disease mutation. This process is called *pregestational* (i.e., before pregnancy) *genetic testing*, or *PGT*.

PGT is subdivided into three main testing categories. The first type, *PGT-A*, screens for how many chromosomes the fetus has. One might assume that this test should be performed on every preimplantation embryo, but PGT-A is offered only in some medical centers where IVF is performed. Its use is a subject of debate within the professional IVF community; many centers do not believe it increases the number of healthy babies born. Another test, called *PGT-SR*, focuses on structural rearrangements of the chromosomes. Last, there is *PGT-M*, which is a test for a specific mutation in the fetus. This can be customized to search for a specific mutation carried by a single parent or by parents who are nonaffected carriers of a disease.

PGT is performed in the lab during day three or day five following fertilization (depending on the lab). An experienced technician aspirates a small number of cells from the developing embryo. The genetic material is then amplified to produce enough DNA to verify the presence of a genetic condition. If the result is good, the embryo can be implanted in the mother's womb. If the result is bad—meaning that technicians find evidence of a genetic abnormality—the embryo is not selected for implantation.

In recent years, a cell-free technique has been developed that involves testing the fluid near the embryo, rather than the embryo itself. Researchers believe this may increase the chances for a successful pregnancy.

Advancing successfully toward embryo implantation can depend on the type of genetic disorder in the parents. For instance, if one of the parents is a carrier for a dominant disorder, 50% of the IVF embryos will potentially be fit for implantation, and 50% will be discarded. If the goal of the testing is to determine whether the fetus has a recessive mutation for which parents are carriers, then 75% of the embryos will be returnable and 25% not. In both situations, the overall chances of bringing a healthy child to term become lower, because of the lower number of fertilized cells available for implantation.

Preimplantation genetic analysis of embryos comes with a cost; it is associated with a reduced IVF success rate. Another, more personal cost stems from the fact that couples who use IVF to avoid genetic disease in their offspring must commit to restricting unprotected intercourse. A pregnancy that begins outside the procedure—that is, a pregnancy that is conceived naturally—could jeopardize the entire process. For many couples, the trade-off feels worthwhile, because it offers a reliable strategy for preventing the transmission of harmful genes. Yet, for others, the allure of unprotected intimacy outweighs the benefits of genetic testing. This leads them to choose the more invasive route, where the risk is contained but the freedom to conceive naturally remains intact.

12.1.8 Who Makes the Decision About testing, the Parents or the Doctor?

Genetic counselors and fertility specialists offer advice and acquaint parents of the range of technologies that can help them complete a pregnancy and give birth to healthy children. But ultimately, the decision about whether to perform genetic tests and what to do when test results are received is made by the parents alone.

Open communication between the parents, the genetic counselor, and the medical team is essential for helping parents reach the decisions that best align with their values.

Chapter 13
Same Syndrome, Different Symptoms, and the Case of the Hidden Disease

When a genetic syndrome is passed down from parent to child, it can often come as a complete surprise. This is true of cases where a mutation generates no obvious physical or cognitive abnormalities in the parent. But sometimes, a parent's problematic genetic profile is hiding in plain sight.

The symptoms of genetic disease can manifest in wildly different ways and with different degrees of severity in various family members. In an adult, very mild symptoms can be a blessing. Unaware of any underlying genetic condition, the individual enjoys a full and healthy life. When this same individual decides to have children, however, the hidden inheritance may become a curse. This happens when it shows up with a vengeance in the next generation.

A child's genetic diagnosis can reveal the presence of previously overlooked symptoms in one or both parents. Although unavoidable, this scenario represents a missed opportunity; had the symptoms been noticed at an early stage, the parents could have been referred to a clinical geneticist or genetic counselor, who would likely have called for the tests needed to ensure the birth of a healthy baby.

But better late than never. Thanks to modern testing techniques, parents can gather the information they need to understand and care for their affected child. Further, the results of genetic tests allow the counselor to guide the parents so that they can adopt a strategy for ensuring that any subsequent pregnancies will result in children born without the syndrome-causing genetic inheritance.

This is what happened in the case of Mike, Isabella, and their daughter Juno.

> During my pediatric internship, I saw many children with neurological disorders. One of these was a 7-month-old baby girl with beautiful blonde hair. Her name was Juno, and her parents, Isabella and Mike, had brought her to the hospital emergency room. Although throughout the discussion, Juno was quiet and observed me calmly through her bright, almond-shaped eyes, her parents were clearly terrified.
>
> "Over the last day or two, our daughter has been having what looks like some kind of physical seizure," Isabella began. "There's this repetitive bending over from the waist, and for as long as it lasts, she seems to retreat into herself."

E. Banne, B.-Z. Shilo, *Dialogue With Our Genes*,
https://doi.org/10.1007/978-981-95-9571-6_13

"It's like we can't wake her up from a bad dream, even though we're holding her and trying to calm her down," Mike added. "The whole thing is over in a few minutes, but it's happened three times now. I know some kids have seizures when they have a high fever, but this is different. Juno wasn't even sick."

Isabella added some important background. "From the time Juno was about 6 weeks old, she started crying for hours every day, and this went on until about a month ago," she said. "I know that we're new parents, but still, we suspected that the crying was something significant, not 'just colic' like our pediatrician told us." Isabella described how at the end of these crying bouts, Juno's body would sometimes stiffen. She wondered if this stiffening had any connection to the seizures.

In addressing the parents' fears, I decided not to address their pediatrician's diagnosis. I know that *infantile colic* is a generalized term for the discomfort babies experience during their first few months of life without a known cause. Although many anticolic treatments are available on the market, they mostly function to make money off distraught parents. Eventually, the problem usually resolves itself.

Since Juno's current spasm episodes were intermittent—and because the child showed no symptoms during her visit to the emergency room—I decided not to offer the worried parents a tentative diagnosis. Instead, I extended emotional support, as well as a simple plan for getting the information needed to move ahead.

"What you're describing sounds like a type of infantile spasm, and it certainly sounds frightening," I said. "You might be correct in your observation, but there is also the chance that there's nothing to worry about. I suggest we admit you to the pediatric ward, just overnight, to get a better idea. If this happens again, I'd like you to make a video and show it to the staff in the morning. Once we see what we're dealing with, we'll order the necessary tests."

Sure enough, the next day during my morning rounds, my cell phone pinged. It was a message from one of the residents. Isabella had sent her a video clip showing Juno in the midst of what doctors sometimes refer to as a *salaam attack*. This is because the repetitive bowing from the waist in this type of seizure is reminiscent of something seen all the time at the Jerusalem hospital where I used to work: Muslims at prayer.

I knew that the family would need genetic testing—not prayer—to discover what was causing their daughter's symptoms.

Infantile spasms are a type of childhood seizure first characterized in 1841 by surgeon and apothecary William James West. After observing an attack in which his own 4-month-old baby suddenly displayed wild and involuntary movements of the limbs and eyes, and loss of consciousness, West described the experience in *The Lancet*, one of the medical world's most venerable medical journals. Since then, *West syndrome* is the name used for a specific subset of the seizure disorders that can affect infants and children.

Today, doctors diagnose infantile spasms based on three things: the visual symptoms observed in the clinic, a blood test to rule out other conditions, and the readout of an electroencephalogram, or EEG, a test that detects electrical activity in the brain via electrodes attached to the scalp. To diagnose a patient suspected of having infantile spasms, doctors search the EEG readout for something called *hypsarrhythmia*, a complete irregularity of the electrical signals in the brain.

Once the diagnosis is confirmed, the patient is treated with adrenocorticotropic hormone (ACTH), a very potent hormone, and/or prednisolone, a more common steroidal hormone. This simpler, cheaper drug causes fewer side effects and can be similarly effective as ACTH and is used alone or in combination with an antiepileptic drug that suppresses convulsions.

As I continued my rounds, I arrived at Juno's hospital room. Again, the child was calm, resting happily in Mike's arms. Examining her, I noticed some dermatological abnormalities. Two spots on her skin were brighter than the surrounding areas. A small patch of skin on her lower back was raised, leathery, and dimpled like an orange peel.

While Juno's parents were aware of these unusual skin characteristics, it had never occurred to them that they might have medical significance. Together with Mike and Isabella, I began the information-gathering process.

"Let's talk about your family background," I began. "Are you aware of any syndromes that may have affected your parents, grandparents, or even your great-grandparents?"

As we discussed in Chap. 6, family trees are a powerful tool that can help focus the search for a symptom-causing genetic mutation. In this case, the parents knew of no family members with medical disorders that might be relevant to Juno's condition. This did not close the door on a diagnosis, however, as infantile spasms are not always genetic.

"Infantile spasms are caused by chaotic signaling in the brain, similar to what happens in epilepsy," I said. "The best way to get to the bottom of this would be to perform a brain scan."

I explained that I would be consulting with our pediatric neurologist and ordering brain imaging using *magnetic resonance imaging* (*MRI*), a test that uses a magnetic field and radio waves to produce detailed images of the brain and brain stem.

"Because the patient has to stay absolutely still, something impossible for a child this age, we'll be giving your daughter a sedative before the scan. She'll sleep right through it," I said. "In addition, I'd like to keep Juno here in the hospital, just to keep an eye on her until the MRI scan gives us the information we need."

They agreed.

When tracking down a genetic syndrome, doctors sometimes arrive at a diagnosis by way of phenotype—the patient's observable characteristics—rather than through an analysis of the patient's genetic profile, or genotype. This was the case with Juno. Her MRI scan revealed a clearly observable physical symptom: small structures in her brain called *tubers*. This finding immediately narrowed the possibilities.

Cortical tubers are benign brain lesions. They are composed of dysfunctional cells that remain undifferentiated, arrested in their development and never committing to maturing into a particular cell type. A condition that sometimes arises during early brain development and can be detected from birth, cortical tubers indicate the presence of a genetic disorder called *tuberous sclerosis complex*, or *TSC*.

Tuberous sclerosis complex is caused by mutations in the TSC1 and TSC2 genes. *Sclerosis* means "hardening of tissue," and tubers are root-shaped, noncancerous growths that develop all over the body. These tubers are seen in the brain of nearly 80% of all TSC patients and can also appear in vital organs like the kidneys. Significantly for Juno and her diagnosis, the condition can also result in raised, hardened patches on the skin. These are called *shagreen patches*, after the stingray skin it resembles.

In addition to this important information about phenotype, I ordered tests that would use genetic sequencing to reveal any genetic abnormalities. When the tests came back, the results were definitive: the child's genotype showed that she had TSC, tuberous sclerosis complex.

Since Juno's observable symptoms pointed to tuberous sclerosis, the ordered tests focused on two genes: TSC1 and TSC2. The test results revealed a mutation in the TSC1 gene. The mutation inactivated one copy of the protein for which this gene codes so that it was unable to perform its proper function.

The test revealed an abnormal substitution in the sequence of nucleotides, the basic building blocks of nucleic acids like DNA and RNA. In Juno's DNA, the nucleotide guanine (G) appeared in the spot where a different nucleotide, adenine (A), would normally be positioned.

This type of substitution does not have any effect at the DNA level. However, when protein synthesis begins, problems arise. What problems? Well, it depends.

Let's take a look at Juno's G-to-A switch, the seemingly minor substitution of one guanine for an adenine. The substitution meant that when the genetic code was translated to make a protein, a different amino acid would be inserted at this position. In Juno's case, the substitution took place in the TSC1 gene, which is responsible for producing a protein called *hamartin*.

Under normal circumstances, hamartin creates a structural complex with another protein called *tuberin*. Once joined together, the hamartin-tuberin complex acts like the brakes of a car, making sure that cell division happens only at the right place and time, and at the required magnitude. Damage to one component of the brake can result in uncontrolled cell division, creating tubers—the benign lesions characteristic of tuberous sclerosis disease. Although, for Juno, only a single amino acid was altered in a total sequence of 1164 amino acids, this seemingly minor change nevertheless made a functional difference, and that difference led to her symptoms.

There are about a million people worldwide with TSC. At least two babies born each day will have TSC, one baby in every 6000 births. Many of those who carry the disease are never referred to a physician; they have only very mild symptoms and never need medical attention. However, in other patients, the symptoms are quite serious. How can this variability of the syndrome be explained? Again, the difference is rooted in dynamics that play out on the genetic level.

The two genes that code for proteins that make up the TSC complex are large and thus represent a large target for modification by mutations. Such mutations may arise from exposure to UV light or toxic chemicals, or simply from a random rare mistake in gene replication. The chance of such mutations occurring in both copies of one of the TSC genes—that is, in the TSC genes received from each of the two parents—is very small. However, people who start off with one defective copy have a heightened chance of developing the syndrome. Any such change will compromise function, by making the tuberin-hamartin protein complex abnormal.

Depending on the circumstances, TSC patients may have impaired protein function in a small number of cells. Such lucky patients, who display a mild phenotype, may need no medical treatment. Other patients have the second mutation in a large number of cells, resulting in a severe case. The tissues in which the disease will be manifested—like the tubers seen in Juno's brain scan and the raised ridges on her skin—depend on where the mutations occurred.

So, as we've learned, tuberous sclerosis is an inherited genetic disease. When a child receives one faulty copy of the TSC gene from a parent, this increases the chances that a mutation will be generated in the remaining normal copy. Further, the occurrence of a second mutation in specific cell populations determines how the disease manifests. This explains the variability of symptoms among TSC patients.

Now, on the levels of both observable phenotype and genetic sequence, I understood what had caused Juno's symptoms. However, the clinical question remained: what was the best treatment?

Like other babies with infantile seizure disorders, Juno was initially treated with a drug called *vigabatrin*. While she had no negative response, her EEG remained abnormal. The steroid prednisolone was added to control her seizures, which could result in permanent neural damage if allowed to continue unchecked. After she spent 2 weeks as a hospital in-patient, Juno's condition was improving. Eventually, the seizures stopped, and the prednisone was slowly discontinued.

With Juno's symptoms under control and the results of her MRI and sequencing in hand, the time had come for me to meet with the parents and prepare them for the challenge of caring for their daughter at home. Together with a senior colleague and a social worker, I met with the family in a conference room.

As Mike and Isabella had been switching off their time in the hospital with Juno, we scheduled the meeting for late afternoon to accommodate Isabella's work schedule. Mike, who had spent the day at his daughter's side, was on time for the meeting, but Isabella was running late. I knew that Isabella was struggling with the logistics associated with Juno's hospitalization, as well as the emotional pressure associated with returning to the workforce after maternity leave. Mike, too, was under pressure from his employer because of all the time he was spending in the hospital and away from the office.

As we waited for Isabella, Mike checked his phone and found a text message from his wife. "She's just parking, now," he said. "I'm sorry."

"Not to worry," I assured him.

When Isabella arrived, I began by informing the parents that all the tests had come back and that they confirmed Juno's seizures were caused by tuberous sclerosis complex, or TSC, a genetic syndrome. Though neither parent was surprised by the news, Mike wanted to understand how this had happened.

"How do you explain the fact that our child has something no one in our family has?" Mike asked, stifling a yawn and brushing his hair away from his face.

It was then that I noticed something I had not seen before: a reddish-brown rash distributed on Mike's nasal bridge and both of his upper cheeks.

"How long have you had a rash on your face?" I asked.

"For as long as I can remember," Mike replied. "Why?"

"Have you tried treating it?"

"I've tried acne creams and other medications, but it doesn't go away."

I paused and then ventured into what I knew would be an unexpected line of questioning.

"Did you ever have a neurological problem?" I asked Mike. "Of any kind? For example, did you ever need special help in school?"

Startled, Mike shook his head no.

"Any kidney or lung problems? Any teeth or gum problems?"

"No," Mike said. "Why do you ask?"

"If I am not mistaken," I answered, "you have adenoma sebaceum on your face."

I explained to Juno's parents that this rash, though unlike Juno's own skin patch, belongs to the same group of skin changes that appear in over 85% of people with tuberous sclerosis.

Mike was astounded to learn that he and his daughter shared the same genetic syndrome. This realization reminded him of something: a stubborn crusty patch on one of his toenails that had persisted for years. On inspection, it was clear that this patch was also part of the syndrome's manifestation, something that was later confirmed through genetic testing.

"Since the tuberous sclerosis genotype has had no serious effect on your day-to-day life, there's nothing to worry about," I assured Mike, "and our first order of business here is to map out a continuing care plan for Juno. All the same, I would advise you and your wife to get genetic counseling for yourselves and schedule a full workup. This can be done whenever you feel you have the time but certainly before your next planned pregnancy."

Despite the mildness of Mike's symptoms, I knew that as parents, he and Isabella had a long road ahead. That's because the phenotype that their now 8-month-old daughter currently displayed could be just the beginning. The benign brain tumors associated with TSC can have significant consequences for neurological development. This and more: while Mike's mild phenotype had no impact on his day-to-day life, complications could still develop. He would have to commit to regular medical imaging of certain internal organs, where life-threatening TSC symptoms might possibly develop.

As someone who had come to care about this young couple and their daughter, I could only hope that skin rashes and seizures, which could be controlled with medication, would be the family's most serious concern going forward.

Chapter 14
Out of the Blue: De Novo Mutations

As we have seen, clinical genetic counseling provides tools for understanding how genetic mutations carried by one or both parents are passed down to offspring. But a child can also be diagnosed with a mutation that is not seen in either of the parents' genetic sequence. Recent research about the provenance of this type of hereditary abnormality—what scientists call de novo *mutations*—has made it possible to pinpoint exactly when such variations first appear in a developing embryo.

Let's take a look at what can cause a mutation to appear seemingly out of the blue.

As you will recall, nucleotides (which we have met before as the bases A, C, G, and T) are the building blocks of nucleic acids, large molecules like DNA that carry the genetic code that produces traits. The genes contained in DNA code for proteins, the biological molecules that carry out tasks within cells. However, the accuracy with which proteins are generated depends on whether their fundamental building blocks, known as *amino acids*, arrive at the right place and line up in the correct order.

But there's more. The gene sequence encoded into DNA is passed on to a form of nucleic acid called *messenger RNA* (*mRNA*). The job of mRNA is to read instructions from the DNA and translate it into a working plan for protein synthesis. Protein synthesis takes place in a cellular organelle—a tiny structure inside the cell—called the *ribosome*. When instructions delivered by the mRNA contains a mutated sequence, the result is faulty; the proteins assembled in the ribosome may not perform their normal function.

To explain how this unfortunate turn of events occurs, we must introduce a new genetic term: the *codon*.

A codon is a set of three nucleotides that, together, form a unit of instruction for protein synthesis. Sixty-four different codons serve as a vocabulary for telling the cell's protein-making machinery to take action. A *start codon* directs the ribosome to begin construction of a protein chain. Other codons call for the addition of a particular amino acid. A *stop codon* terminates protein synthesis.

E. Banne, B.-Z. Shilo, *Dialogue With Our Genes*,
https://doi.org/10.1007/978-981-95-9571-6_14

Any change in the DNA sequence will affect how the codons are read as well as how the message is understood. In some cases, this will lead to abnormal protein production. Let's look at a linguistic example that illustrates the concept.

Consider the sentence "The cat ate the red bat." Note that this sentence contains words made up of just three letters. Now, let's generate a mutation in this sentence, by deleting *t* from the word *ate*. The sentence will now read: "The cat aet her edb at." This change—the deletion of a single letter—turns the word and the rest of the sentence into gibberish.

This sort of thing can also occur on the genetic level, triggered by the deletion of a single nucleotide. Without this genetic letter, the word containing instructions for building an amino acid is read incorrectly. What's more, the lack of a letter causes every subsequent amino acid in the sequence to be nudged out of its normal position, a phenomenon that scientists call a *genetic frameshift*.

Frameshift can appear in a child whose genetic inheritance looks completely normal and whose parents' genes look normal as well. In such cases, the mutational frameshift occurred in germline cells that give rise to the parents' egg or sperm prior to fertilization. This causes the developing embryo, and eventually the child, to carry DNA that is ever-so-slightly different from what would be expected. A tiny change with potentially large consequences.

To see how the tiny change that leads to genetic frameshift can have major repercussions, let's now go back to Dr. Banne's genetic counseling clinic. We'll also go back to a family we met at the very beginning of this book: Shiri, a sweet little girl with developmental delay; her parents, Keren and Yoav; and Professor Benny Shilo, the expert in genetic science who arrived at the clinic in a different role, that of Shiri's worried grandfather. Again, we'll join the discussion from Dr. Banne's point of view.

"Good to see you again," I began. "Today, we'll be discussing what we can see in the sequencing done on Shiri, as well as both parents."

Although this family was better informed about genetic science than most of my patients, I carefully laid out the story step by step. "As you know, based on Shiri's medical exam, no genes could be identified as the likely cause of her condition. That's why I asked for exome sequencing, an analytic technique that covers the entirety of the genetic material that codes for proteins, meaning about 1% of the total DNA."

I explained that when Shiri's genes were compared to those of her parents, one copy of a particular gene couldn't be traced to either her mother or her father.

"Since a child receives one copy of each gene from each parent, we would usually expect the child's sequence to match up." I considered adding what I generally tell my patients by way of clarification, that the two copies of each gene that a child carries should be directly traceable to one-half of the genetic material carried by each parent. In this case, however, I felt that the whole family was familiar enough with genetic science that the concept was well understood.

After making eye contact with Keren and Yoav, then with Benny, I continued. "In Shiri's sequence, we see that one of her genes is different. A gene called *SHANK2* is missing a single nucleotide: the *C* for cytosine."

A loud silence filled the room, as the parents and the professor took in this information.

"The missing nucleotide appears at the position in the sequence where the 1384 amino acid is encoded," I explained, adding for the parents' benefit that amino acid positions are identified by numbers. "The change means that not only is amino acid 1384 affected but everything in the continuation of the sequence—right up to the point where a stop code marks the end of a protein—becomes garbled. This causes frameshift. From the point of the deletion on, the wrong amino acids are inserted into the SHANK2 protein. This change in a single nucleotide affects the SHANK2 protein's function."

"And what is SHANK2's function?" Keren asked.

I kept my voice low and steady, aware that what I was about to say would be difficult for the family to hear. "SHANK2, and a related gene called *SHANK3*, are important for the proper development of the nervous system."

The nervous system is a complex network of organs, tissues, and cells that orchestrates communication between different parts of the body. At the heart of this system are *neurons*, the specialized cells that transmit information using electrical and chemical signals.

Neurons come in many different shapes and sizes, but they all share some common features. Each neuron has a cell body, which contains the nucleus and other organelles necessary for maintaining the cell's functions. Extending from the cell body are long, branching structures called *dendrites*, which receive signals from other neurons. At the other end of the neuron is the *axon*, a long, thin projection that transmits signals to other cells. Together, dendrites and axons enable a neuron to form networks and to communicate with other neurons.

Neural communication occurs at specialized junctions called *synapses*. When a neuron is stimulated, it releases neurotransmitters, chemical messengers that travel across the synapse to bind to receptors on the surface of the receiving neuron. This triggers a cascade of events that ultimately results in an electrical impulse being generated in the receiving cell, allowing the signal to be transmitted along the neural network.

Although synapses are tiny, they are incredibly complex. A multitude of proteins and other molecules work together to ensure proper synaptic function. Disruptions in this function are associated with a range of neurological disorders, including epilepsy as well as ataxia, a rare progressive disease that affects a person's ability to walk, talk, and take actions that depend on fine motor skills.

The SHANK2 protein is critical for cross-synapse communication. It acts as a structural scaffold, helping to anchor other proteins to the receiving end of the synapse. By doing so, it ensures that neurons can successfully receive and process chemical signals.

Recent studies have shown that disruptions in **SHANK2** production can have significant implications for brain function. In some cases, mutations in the **SHANK2** gene have been linked to a range of neuropsychiatric conditions, including autism, schizophrenia, and bipolar disorder.

How do lower levels of **SHANK2** affect the nervous system?

Imagine each **SHANK2** protein as a crossbar that strengthens the kind of scaffold used at construction sites. If there are not enough stabilizing crossbars available, the entire structure may collapse. Even if the scaffolding continues to stand, an unsteady platform will compromise the construction worker's performance.

This is what happens in the case of a mutation involving a single-nucleotide deletion that affects **SHANK2**. The transfer of signals is compromised because the structures that receive the signals are wobbly and unreliable.

Insufficient levels of **SHANK2** apparently reduce the nervous system's ability to establish and strengthen frequently used synaptic connections—a fundamental dynamic associated with learning and memory—and lead to a more dispersed network of neurons. This phenomenon, believed to be mediated by **SHANK2** proteins active during the second trimester of pregnancy, affects fetal brain development, structure, and neural connectivity on the most basic level.

As I shared the news about Shiri's diagnosis, Benny stayed mostly quiet, preferring to let his daughter and son-in-law direct the conversation. But he later found the opportunity to discuss the case privately.

"It's ironic to see how things I have been grappling with throughout my scientific career—how genes and traits are passed down—have now shown up in a way that affects my life," Benny said. "Of course, I've been aware of my granddaughter's developmental delay for some time, but this news puts it in the context of genetic mechanisms that have been conserved in evolution. What went wrong at some point during Shiri's development involves the same mechanism that stabilizes synapse formation in every multicellular organism with a nervous system."

As much as I shared this professor's passion for science, I understood Shiri's diagnosis had a personal impact as well. "How did getting this news make you feel?" I asked.

"Well, my initial response was shock," Benny replied. "It was hard to get a diagnosis that unequivocally placed someone I love on the autistic spectrum. Over time, however, this information has become valuable."

"How so?"

"It allowed my granddaughter to be placed in a special kindergarten suited to her needs," Professor Shilo said. "We were able to get her into this framework because her genetic condition was identified early."

Benny paused before adding, "The genetic diagnosis also allows me to view Shiri in a different way. Because I understand why she is different, I am now less likely to fall into the trap of pressuring her to behave in ways that are not appropriate to her. I'm not saying that it's always easy, but it does help me accept and love her as she is."

Parents of children who have already been diagnosed on the autistic spectrum, like Shiri, often ask: why carry out a genetic analysis, if the medical diagnosis is already known?

In the case of autism, this is not an unreasonable question. Only about 5% of autism cases are associated with a mutation in a single gene, while the rest are caused by a combination of multiple genetic alterations that are difficult to pinpoint. However, in situations where a single mutation is identified, as was the case with Shiri, genetic testing can be valuable in several ways.

Linking a behavioral syndrome with a specific genetic diagnosis can have beneficial psychological ramifications. It can help parents accept the situation and quickly seek the most suitable educational framework. It can enable them to connect with families that face similar challenges and learn more about what to expect and how to best meet their child's needs at home. In cases where a mutation is identified, this information can be used for the prenatal diagnosis of future children. Finally, evidence from genetic testing can entitle the family to benefits that a more general diagnosis may not provide, such as tax reductions or disability assistance, if they choose to reveal it.

14.1 Out of the Blue

We now return to the second open question described at the beginning of this chapter. How is it that a genetic mutation that is not seen in either one of the parents appears in the child? Where does such a mutation come from, and when does it happen?

The exome sequencing performed on Shiri and her parents analyzed the genetic material contained in somatic cells—that is, body cells that are not involved in reproduction. However, the mutation was not identified in the somatic cells of either parent. Instead, this mutation originated in one of the parents' germline cells, precursors destined to mature into egg or sperm. Faulty cell division that leads to the creation of either an egg or sperm cell containing an incorrect nucleic acid is called a de novo *mutation*.

A mystery remains: did Shiri's de novo mutation originate in her father or her mother? Males produce sperm cells continuously, so if Shiri's mutation came from Yoav's sperm, it likely occurred recently. Women, on the other hand, are born with all their eggs. Therefore, if the mutation came from Keren's eggs, it could potentially be traced back to faulty cell division that occurred when Keren was still in her mother's womb, sometime between the second and fifth week of her own embryonic development.

Awareness of de novo mutations blossomed in the late twentieth and early twenty-first century, as DNA sequencing became a widely used tool. Then, in 2021, a landmark study by researchers in Iceland succeeded in answering two related questions: at what stage does a mutation found either in the mother's egg or the father's sperm appear, and at what stage does this lead to a de novo mutation in the parents' offspring?

In the study, a team of scientists from Iceland examined identical twins - who originated from the same sperm and egg. The scientists sequenced the genes of the twins' parents, the twins themselves, and the twins' own offspring. Although identical twins were long believed to have identical DNA as a rule, the Icelandic researchers demonstrated that twins can have genetic differences due to mutations that occur during early development.

The scientists found three types of de novo mutations: mutations present in both twins but absent in their parents, indicating they originated in a parent's reproductive cells; mutations found in only one twin, suggesting they occurred after the embryo split; and mutations appearing in the children of one twin but not in the twin's own body cells, meaning they arose in early embryonic development in the cells that later became sperm or eggs.

These findings challenge the assumption that all differences between identical twins are purely environmental and a product of the circumstances in which they were raised. They suggest that not all differences between identical twins can be explained as a matter of nature versus nurture. Instead, genetic mutations occurring very early can influence traits as well as disease risks.

"So how do you advise parents to proceed after they've learned that one of them carries a germline mutation?" Benny asked. "After all, this is something that couldn't have been known until after the child is born."

"In cases like this, it is important for the counselor to make sure parents understand that their child's condition is not their fault," I responded. "Then, we can move on to the practical, clinical goal: giving parents the tools they need to avoid passing this same mutation to additional children. For example, we can suggest that during subsequent pregnancies, the parents use amniocentesis or chorionic villus sampling to see if their fetus received the same mutation. If the developing fetus is found to carry the mutation, the parents may elect to have an abortion. These are hard choices, but the goal of genetic counseling is to do everything we can to make sure parents have the information they need to make their own decision." For families like Shiri's, understanding that a mutation arose 'out of the blue' does not erase the challenge - but it can replace guilt with clarity.

Chapter 15
The Chromosomal Origins of Sex-Specific Disease

We'd all like to believe that when we pass down our genes, we're giving our children an equal shot at a healthy life. Unfortunately, our chromosomes conspire against this. Females carry two X chromosomes, while males carry one X chromosome and one Y chromosome. This fundamental difference generates some sex-specific and gender-specific health outcomes. When parents' genes recombine and assemble into the chromosomes found in every one of a developing baby's cells, boys inherit a built-in disadvantage: they have a greater risk of being born with certain genetic diseases.

What happens during chromosome formation that puts our boys at risk? The Y chromosome, found exclusively in males, is shorter than the X and holds less genetic information. Boys' heightened risk for genetic disease, however, isn't rooted in anything missing in the Y chromosome as opposed to the X. On the contrary, although Y chromosomes are short, they play a significant role in development, carrying critical genetic information. This includes the transcription factor SRY that directs sexual differentiation and the sex determination of both boys and girls.

Sex-specific genetic syndromes that primarily affect males include hemophilia and Duchenne muscular dystrophy and less dire diagnoses like male pattern baldness and red-green color blindness. Ironically, these syndromes stem from mutated genes located on males' single X chromosome rather than the Y.

What is it about the location of mutated genes on the X chromosome that increases the likelihood of developing a genetic disorder? To answer this question, we need to go back and review the laws of Mendelian genetics.

As you may recall, we all carry two copies of each gene. Inherited traits are encoded in a single gene, but that gene can appear in one of two alleles, or versions, one version of the gene inherited from the mother and the other from the father. Some inherited traits are *recessive*, meaning that it will only appear when an individual inherits two copies of the recessive gene, one from each of the parents. Other traits are *dominant*, meaning that it will appear in an individual even if only one

E. Banne, B.-Z. Shilo, *Dialogue With Our Genes*,
https://doi.org/10.1007/978-981-95-9571-6_15

copy of the gene for that trait is inherited. A dominant allele masks the effect of a recessive allele if both are present.

When a girl inherits a recessive mutation from either her mother or father and when the mutated gene appears on one of her X chromosomes, there is a 50% chance that expression of the trait will be blocked. If the gene is recessive, the tissue will compensate by using the nonmutated version, and the tissue will not express disease. If the mutated gene inherited from one or both parents is dominant, the disease-associated trait will be expressed.

Consider now the same situation, but in a boy. A gene coding for a disease-associated trait is inherited from one parent. When that gene is located on the boy's single X chromosome, there is a 100% chance that that trait will be expressed. There is no second allele to potentially override expression of the trait. According to the rules of Mendelian genetics, if the inherited gene is dominant, the trait will appear. If the gene is recessive, the trait will also appear, because there is no second X chromosome capable of carrying a second version of the gene to block expression of the trait. In cases involving a disease-associated mutation, this dynamic can lead to severe and even life-threatening developmental defects seen overwhelmingly in boys rather than in girls.

Duchenne muscular dystrophy is an example of how a mutated gene on the X chromosome can lead to disease symptoms that principally affect boys as opposed to girls. Duchenne is an incurable condition in which muscles weaken and break down over time. A minority of girls who inherit the Duchenne mutation may have some muscle weakness and can also pass down the condition to their offspring. But they rarely experience the disease's full effects the way boys do.

Usually diagnosed in early childhood, Duchenne symptoms include difficulty walking, climbing stairs, and getting up from the floor. Children with this disease may appear clumsy and tend to fall more often than others. By their teenage years, most Duchenne patients rely on wheelchairs for mobility, and as they age, the disorder can impact the heart and lungs. The primary treatment involves steroids, which carry a long list of adverse effects, presenting significant challenges in this therapy.

Duchenne is caused by a mutation in a gene that maintains muscle strength. Therefore, one might hope that gene therapy—a collection of methods focused on transplanting normal genes into cells to replace or supplement genes that are missing or defective—could provide a strategy for effective treatment.

Unfortunately, in the case of Duchenne, this presents a significant challenge. The critical gene that codes for this condition is enormous in molecular terms, making it technically challenging to create normal copies of the gene using current technologies and place them in a carrier (also known as a *vector*). There simply isn't a carrier of sufficient size available.

To work around these limitations, some medications are designed to minimize the damage Duchenne can cause. They may do so by targeting the mutated gene so that it generates proteins that are smaller and less harmful to muscle function. These shorter proteins preserve muscle for more years, making the disease milder. Other medications can bypass stop codons and replace the steroids typically used to treat this condition.

15.1 The Delicate Balance of XX Gene Expression

The living cell is a bustling factory, where complex networks of proteins orchestrate everything from cell growth and division to the timing of a cell's ultimate breakdown and death. Many proteins work together in these networks, and each protein must be available in the right amount. Therefore, a question arises: does the presence of two X chromosomes in females mean that girls' genes produce twice the number of proteins produced by boys?

In short, the answer is no. Over the course of evolution, a mechanism evolved, ensuring that females produce the same level of X-encoded proteins as males. This critical mechanism is called *X inactivation* or *lyonization*, in honor of Mary F. Lyon, the British geneticist who discovered this gene-silencing phenomenon in the early 1960s.

Lyonization is based on RNA, the single-stranded nucleic acid chains that help convert the information contained in the cell's DNA into functional gene products.

One type of RNA plays a starring role in the lyonization process. Produced by the XIST gene during early embryonic development, this RNA is long and noncoding—that is, it is not involved in the production of any functional protein. Instead, *XIST RNA* coats the X chromosome from which it is transcribed, effectively silencing that chromosome and preventing the excess protein production that would otherwise occur. Which X chromosome is silenced during lyonization is a choice made randomly at the level of each individual cell.

Lyonization explains why women, who have two X chromosomes, don't produce twice the amount of X proteins seen in men. It also explains why men can manage with only one X: to perform most cellular functions, our cells don't need the proteins produced by more than a single X chromosome.

This is not to say that females' extra X is not important. Between 10% and 30% of the genes on this second X chromosome are not silenced. And a good thing too; these genes are associated with traits that require both alleles to function normally.

15.2 Syndromes That Skew Male or Female

As we have seen, it is vital for protein balance and normal cellular function that one of the two X chromosomes healthy females have is inactivated. But sometimes this process goes awry. This can lead to genetic diseases diagnosed primarily in females or primarily in males.

Turner syndrome is a female-only condition that affects about 1 in every 2000 baby girls. Girls born with Turner syndrome have only one X chromosome instead of the usual two. Because of this, they produce smaller amounts of certain proteins that normally come from genes on the second X chromosome—specifically, those few genes that usually remain active even after one X is silenced in typical females. They will also express X-linked disorders in the same manner as males, since they do not have the second X chromosome to protect them. This reduced protein production leads to an abnormal phenotype. Turner syndrome sufferers are short, have a wide or webbed neck, and may have cardiac defects. They are also infertile due to malformation of the genital system. This is an example of females expressing X-linked disorders in the same way as males.

Other X-linked syndromes appear when a baby girl is born with extra X chromosomes in addition to the normal two. While rare, such syndromes—including triple X syndrome (three X chromosomes), tetrasomy X (four X chromosomes), and pentasomy X (five X chromosomes)—are associated with a range of physical and intellectual disabilities.

Some X-linked diseases primarily target males. In these cases, males bear the brunt of disease phenotypes, while women may show mild symptoms or none at all. One genetic syndrome found predominantly in men is known as *Fragile X*, and it is common cause of intellectual disability in males.

Under certain staining methods, the chromosomes of Fragile X patients seem abnormal when viewed under a microscope. The staining reveals gaps—regions prone to genomic instability. Over 80 fragile sites have been identified, and some are linked to genetic disorders. One such site, located on the X chromosome, correlates with Fragile X syndrome.

15.3 Caution: Extreme Imbalance

As mentioned above, the choice of which X chromosome is silenced occurs in every cell that has two X chromosomes, and that choice is random. Now, imagine that this cellular coin toss comes up heads in over 90% of a developing child's cells. This is what happens in a subpopulation of female-dominant genetic syndromes.

When lyonization is skewed and silences the same X in most cells, serious health consequences can occur, especially if the remaining active chromosome carries a disease-associated mutation.

While Fragile X syndrome is typically more severe in males, it can also affect females. Due to X chromosome inactivation, females usually have milder symptoms, but many still exhibit cognitive, emotional, or physical manifestations. In rare cases of extremely skewed X inactivation where the healthy X chromosome is preferentially silenced, women may experience symptoms as severe as those seen in males.

Another disease associated with extreme skewing that affects primarily females is *Rett syndrome*, a neurological disorder caused by a mutated MECP2 gene on the X chromosome. If the X chromosome that bears this mutation remains active, it generates mutant proteins in most cells, and disease symptoms become severe. On the other hand, if lyonization preferentially silences the X that bears the mutation, patients may be mildly affected or even asymptomatic.

15.4 Considerations for Genetic Counseling

X-linked genetic syndromes can have serious health consequences, and that's where genetic counseling can help.

In male-restricted X-linked disorders, following the family tree is important. It can identify asymptomatic women who may have passed on X-linked recessive disorders to their sons. In cases where a female-restricted disorder is suspected, a complete family tree can identify female relatives who have been diagnosed with a genetic disease. Determining whether women with a clear genetic diagnosis have passed the syndrome on to the next generation, however, can be difficult. Most women with conditions like Rett syndrome do not have children.

In cases where a genetic syndrome is not accompanied by obvious symptoms, clinical diagnosis can require extensive genetic testing, which can take time. While waiting for the lab results to return, counselors can give parents important guidance. Patterns in the family tree may prompt the genetic counselor to raise the possibility of informing the patient's extended family—with consent—making them aware that children born in their bloodline, especially boys, may be in danger of an X-linked disorder. This information cannot completely prevent or cure genetic syndromes, but it does give parents the power to minimize risk by making informed family planning decisions.

For women known to be carriers of fragile X syndrome or Duchenne, genetic testing should be part of standard medical practice. The testing can be done during pregnancy or in preimplantation molecular genetic testing (PGT-M) for couples trying to get pregnant through in vitro fertilization. Couples can choose the method that best matches their beliefs, needs, and preferences.

15.5 Parental Preference

What about parents who want to pursue genetic testing even in the absence of a family history of X-related syndromes? Should this option be open to parents? Should such testing be incorporated into standard prenatal treatment for everyone?

In some cases, going along with parental preference could open a problematic ethical door. As some genetic disorders are more common in males, some couples pursuing IVF may choose to carry only female fetuses, even in the absence of a clear genetic diagnosis generated by pregestational testing. This fact on the ground may lead to the problematic request by parents to select the sex of fetuses, even without any identifiable problem in the family.

The request to choose a child's sex may be based on religious belief or may be simply parental preference. But it also enables a form of gender-based discrimination. Giving parents free rein to choose the sex of their child may reinforce harmful gender stereotypes and contribute to societal imbalances. Skewed sex ratios, which have been observed in some countries where male children are culturally preferred, can lead to widespread societal unrest.

Sex selection can be considered a slippery slope leading to eugenics, a controversial ideology that promotes policies designed to limit the birth of children with undesirable traits. A gene-empowered medical technology could potentially allow parents to create "designer babies" by selecting traits such as intelligence or physical appearance. This could exacerbate social inequalities, because only those parents who can afford such technologies would be able to access them.

Sex selection raises questions about the child's right to an open future. Choosing a child's sex based on parental preference might limit the child's autonomy, imposing expectations on them before they are even born.

Is it possible? Yes, it is. As we look to the future, we recognize the need for greater awareness of genetic counseling's power to improve lives while taking care to establish the regulations needed to prevent ethical breaches. These ethical issues are complex, so complex that we have devoted a full chapter to them later in this book.

Our discussion of X-linked diseases raises many medical questions and dilemmas. It is time for another coffee break with Benny.

Ehud: When looking at the molecular biology of X-linked diseases, one feature that I find surprising is the capacity to shut down expression from an entire chromosome. Can you tell me more about how this works?

Benny: Sure, but before I do, it's interesting to note that other species solve the problem of X imbalance differently. For example, in male flies, every gene on the single X they carry is expressed at a level double that in females. This achieves the balanced protein expression that leads to flies' proper biological function.

In mammals, one of the X chromosomes in females is coated by *XIST RNA*. This RNA influences the activity of proteins called *histones*, which coat the chromosomal DNA. The coating increases the propensity of these histones to shut down gene expression.

Ehud: I find it amazing that every cell chooses to coat only one of its X chromosomes, leaving the other X active. When does this happen?

Benny: The cells take this decision early on during the first month of pregnancy because achieving the proper balance is important down the road, during more advanced embryonic development.

Ehud: I am also puzzled by the ability of cells to randomly choose only one X for inactivation.

Benny: As far as we know, this is indeed a random decision, such that half of the female cells will express one X and the other half will express the other. But think about what happens afterward. Since inactivation occurs early in embryo formation, each cell will undergo many divisions. The pattern of these divisions will dictate the distribution of one cell type versus another.

This random dynamic can produce anything from a complete "salt-and-pepper" mix—where both cell types are evenly interspersed—to a pattern where one cell type dominates a large region. If neither the chromosome that remains active nor the one that is shut down carries problematic mutations, the resulting ratio between cell types may not matter. But if one of the X chromosomes carries a mutation in a gene that controls, for example, the proper folding, organization, and movement of brain cells, then having large clusters of cells carrying the mutated gene can lead to disastrous consequences.

Ehud: Right. This is why clinicians need to closely examine family trees and proactively search for possible X-linked conditions. My rule of thumb is to remain cautious and advise patient families to postpone plans for family expansion until we complete the genetic work-up. The possibility of an X-linked disorder can significantly elevate the calculated risk that a child will be born with a genetic condition.

Chapter 16
The Genetics of Gender: Genes, Hormones, and Embryonic Sexual Development

Late in her pregnancy, my wife came to meet me at the hospital at the end of my workday. As she entered the clinic, an older woman, a member of the hospital's cleaning staff, greeted her with a broad smile.

"I can see you're having a boy!" she pronounced. "From the way you're carrying, it's clear as day. May God bless you with an easy labor and a healthy son."

Understanding that there was no arguing with the woman and not wishing to embarrass her—we knew we were expecting a daughter—my wife caught my eye, turned to the woman, and simply said, "Thank you."

As amused as I was by the woman's certitude, I couldn't help but connect this experience to a challenge that comes up in my clinical practice. Of all the diagnostic problems I deal with, some of the most perplexing involve developmental syndromes of sex determination. These conditions, rooted in the earliest days of pregnancy, can sometimes make a person's biological sex less than obvious.

Many expectant parents have strong feelings about the sex of their unborn child. No one would deny the excitement of hearing for the first time—usually during a routine ultrasound—that the mother is carrying either a girl or a boy. However, not everyone is aware of what governs the timing of such an announcement. The fact is, until about the seventh week of pregnancy, the fetus has the potential to go either way.

Why is this the case, and how does this open window for sex determination result in what we all take for granted: a 50–50 chance that any unborn child will be either a boy or a girl? The answer is in genetics or, more specifically, in the sex chromosomes passed down from our parents.

Mammals have a pair of sex chromosomes, one passed down from mom and one from dad. A pair of X chromosomes gives us a girl, and an XY pairing gives us a boy. Since adult females have only X chromosomes to pass on in their egg cells and males can provide either an X or a Y in their sperm, sex is determined by the chromosomal contribution of the father.

But the presence of a Y chromosome is just the first step in the journey to sex determination. To continue to develop as a male, a fertilized egg carrying XY chromosomes must be exposed to very specific chemical stimulation. This stimulation comes from the expression of a single gene. The gene is called *SRY*, so named because it is located on the sex-determining region of the Y chromosome.

SRY encodes a transcription factor that activates a defined array of other genes, directing the development of male structures in the embryo and suppressing the

E. Banne, B.-Z. Shilo, *Dialogue With Our Genes*,
https://doi.org/10.1007/978-981-95-9571-6_16

formation of female structures. Without a functional SRY gene, which only exists in males, in most cases, the embryo develops along the female pathway.

When fertilized eggs carry two X chromosomes, a different type of signaling occurs. A gene called *WNT4* codes for a protein that promotes female sex development while repressing the male phenotype.

Once the genetic signaling is in place, a sex-specific journey begins. To help readers follow this journey, we will need to introduce a few terms associated with the everchanging identity of embryonic cells as they advance toward sex determination. We will also mention some specialized sex-related structures that appear in the fetus in formation.

Primordial germ cells (*PGCs*) are undifferentiated stem cells that give rise to both eggs and sperm. PGCs migrate to the genital ridges, the region of the developing embryo where male or female sex organs will eventually emerge.

Around the fourth to sixth week of pregnancy, both XX and XY embryos have two bilateral sets of tubes known respectively as *Müllerian ducts* and *Wolffian ducts*. Müllerian ducts are involved in the formation of the female-specific structures of the uterus, fallopian tubes, and cervix; the Wolffian ducts promote the formation of male genitalia. One of these sets of ducts will eventually be suppressed.

Before the suppression of either the Müllerian ducts or the Wolffian ducts, the embryo features undifferentiated gonads, precursor tissues that will develop into testes in the male and ovaries in the female. Gonads are responsible for producing the sperm and ova. They also secrete hormones.

By about the seventh week, sexual differentiation moves into high gear with the help of hormones secreted by the gonads. In XY embryos, the presence of the SRY gene and testosterone secreted from the testes triggers the formation of male accessory sex organs that will, in the future, direct the path of the sperm. Under the influence of a testes-derived hormone known as *anti-Müllerian hormone* or *AMH*, the Müllerian ducts disappear. This hormone ensures that males don't develop female sex organs.

16.1 The Hormone Tag Team

As we've seen, both XX and XY embryos initially have the potential to result in a female child, but the chemical message carried by the hormones testosterone and AMH tips XY embryonic development toward masculinity. In the developing female, the key hormones are estrogen and progesterone. In both males and females, a long list of additional hormones plays less central, but still important, roles in sexual differentiation. How are these chemical messages generated, where do they go, and how are they received?

To illustrate how most hormones function on the molecular level, let's imagine sexual differentiation as a relay race. The batons that hormones carry as they run this race are chemical messages passed off at critical locations and points in time.

Like participants in any team sport, all hormones are not created equal. Together, however, they get the job done.

Certain hormones—let's call them *captains*—carry chemical information that instructs other molecules what to do. Captain hormones can be either *peptide hormones* or *steroid hormones*.

Peptide hormones—insulin being a well-known example—never enter the cell. Instead, they bind to receptors on its surface, where the "captains" relay their instructions and set off a cascade of protein activations. This ultimately results in the baton being passed to another class of molecules called *transcription factors*.

As opposed to peptide hormones, sex hormones, like male testosterone and female estrogens, are classified as steroids. To extend our sporting metaphor, steroids don't need to wait for a captain's instructions. They can run the race all by themselves, passing through the cellular membrane and going directly into the nucleus. Once there, they interact with transcription factors, providing protein-building instructions. They recognize precise nucleotide sequences and bind to specific DNA sites, where they gracefully deliver instructions that influence gene expression and regulate the protein production rate.

Transcription factors control the rate at which new proteins are produced and have various ways of regulating the genes that code for proteins. They can regulate them directly or indirectly by using *coactivators* and *corepressors* that change the way the DNA is packaged. As a result of these dynamics, the number of proteins produced is tuned to an optimal level.

Transcription factors do much more than pass on chemical information between runners. They enter the cell and then enter the nucleus, where they scan the cell's genetic material. Based on their ability to recognize precise nucleotide sequences, they then bind to DNA at the specific location to which information needs to be delivered. Once delivery occurs, the chemical message activates or represses particular genes.

Mutations in any of these transcription factors can disrupt normal genital formation, leading to mild-to-serious abnormalities. Mild conditions are usually corrected with surgery during early childhood. In more severe cases, this sort of mutation can result in situations where a person's genetic sex cannot be determined without laboratory testing to determine whether a child carries the sex chromosomes XX or XY.

16.2 Differences in Fetal Sexual Development

Up until this point, we've mostly concentrated on sex determination as it normally occurs. Now, let's take a closer look at how genetic mutations can cause a developing embryo to emerge in the grey area between male and female. This range of conditions is referred to *as disorders of sex determination and differentiation*, or *DSD*.

In some cases, a visual examination at birth shows an atypical development of external sexual organs. Although it is startling for the parents, following surgical treatment, the child can be expected to develop normally, except where fertility is compromised. In such situations, psychological treatment may be needed for the parents and eventually also for the child. Such issues are beyond the scope of this book.

On the other hand, variations in sexual development can sometimes go unnoticed for quite a while. That's what happened in the case of a young girl who was treated in hospital and was later referred to Dr. Banne's genetic counseling clinic.

I first heard about the patient from the resident who received her in the emergency room.

"Eleanor is a 12-year-old girl who arrived with her parents, complaining of severe abdominal pain," the resident said. "She had low blood pressure that administration of fluids failed to bring up, indicating adrenal insufficiency. After stabilizing her with hydrocortisone, we ordered blood tests and imaging. The ultrasound results just came back, and I've never seen anything like it: this patient has no female sexual organs at all—the uterus, fallopian tubes, and ovaries are simply not there. What is the procedure in a case like this?"

In a busy hospital, it is not uncommon for a resident to conduct hallway consultations with experienced physicians, then quickly return to their patients. But from the resident's report, I knew that a more serious investigation was in order. We would need to determine whether this was a case of sex reversal associated with DSD, a genetic disorder of sexual determination and differentiation.

Sex reversal occurs in approximately 1 in 4000 births. As opposed to other, rarer disorders, this may seem relatively common. DSD's prevalence is rooted in the complex molecular pathway that determines sexual development, in which mutations in any of the genetic factors involved can create an imbalance that throws the sexual differential process off track. As previously explained, the SRY gene found on the Y chromosome plays a critical role in transforming early-stage embryos—all of which have the potential to develop female sex organs—into a male. The absence of female sex organs in this case indicates that for this patient, the journey toward a male genetic identity had begun but had never been completed.

Confirming such a diagnosis requires a series of clinical examinations as well as genetic tests. It cannot be done alone; a team of experts is needed. It also requires extreme sensitivity. We knew that whatever the final diagnosis, the patient and her parents were in for a shock.

I turned to the resident. "Before I answer your question, let's set up a joint meeting with the head of the endocrinology unit," I told her, referring to our most senior physician specializing in the diagnosis and treatment of hormone-related diseases and conditions. "I'll call you in when we're both available."

A few hours later, I met with the resident and the endocrinologist in a private conference room. "The first step is to karyotype the patient," I began, reminding the resident, who had little experience in genetic syndromes, that karyotyping was a relatively simple blood test aimed at visually mapping the patient's chromosomes. From that point, my endocrinologist colleague took up the story.

"If the karyotype shows clearly that the patient has two X chromosomes, then she's a genetic female whose sexual organs did not develop. This is likely because of gene mutations affecting the production of hormones that guide the formation of female organs," she said. "Alternatively, if we discover that your patient has one X chromosome and one Y chromosome in each cell, the pattern normally found in males, despite the fact that she has no clearly formed male sex organs, we may be looking at a disorder of sexual development called *46 XY DSD*, which can be triggered by insufficient testosterone production."

I knew that if diagnosed as 46 XY DSD, the patient had a long road ahead, most likely involving surgery and hormone replacement therapy. Individuals diagnosed with this condition also have an increased risk for gonadal tumors, something that would require regular surveillance.

But focusing on treatment at this point would be getting ahead of ourselves. Our first task as doctors was to assemble the proper team for informing the patient, and the patient's family, about what the scan had revealed. We needed to share this information in a way that was both medically accurate and sensitive to the emotional ramifications our findings would inevitably trigger.

"I'll book a conference room, and we'll begin with just the parents," I said. "As this is a genetic syndrome, I'll lead the discussion, alongside the endocrinologist. But it's important that the attending doctor, who is familiar to the family, be present as well. We'll also invite the head ward nurse and a social worker, because there is no question that this news is going to be very difficult to hear. The goal of this meeting is to break the news, then to give the parents the time they need to ask any questions, particularly about how and when to inform their child of the situation. We'll also be informing them about the range of support services available."

There is enormous variability in how DSD presents on the physical level, but whatever the phenotype, individuals diagnosed with this type of disorder are at high risk for mental health problems. They may suffer from traumatic treatment experiences and feelings of being abnormal, causing their lives to be suffused with a sense of secrecy and shame. They may worry about long-term effects, including infertility. Some individuals with DSD report that they experience gender dysphoria and do not feel entirely male or female; such feelings are often associated with lower self-esteem, negative body image, and difficulty establishing a satisfying sex life.

A few days after the patient and her family heard the news, they returned to the hospital for what would be the first of many visits to my genetic clinic.

The abdominal pain that had triggered Eleanor's visit to the emergency room had worn off. A gastroenterological exam showed it was the result of a temporary bowel obstruction, which passed spontaneously. The shock of the ultrasound results, however, did not seem to have diminished; all three showed signs of severe emotional distress. The mother, Denise, was the first to speak.

"So we have a complicated situation on our hands," she said.

"I would say we are at the beginning of a process that will determine what exactly the situation is, which will allow us to determine the best plan going forward," I began and then turned to Eleanor.

"There is no question that discovering something you didn't know about your own body, something that was hidden from you, from your parents, and even from your doctors, can be very stressful," I said. "But I'd like to suggest that the intestinal pain you experienced was a gift: it brought you here so that we can give you the help you need."

My interview revealed that Eleanor—a slim 12-year-old who showed no outward signs of puberty—had not yet begun menstruating. A competent family doctor or pediatrician might eventually have diagnosed this as primary ovarian insufficiency. But these professionals would not have had access to the tools needed to identify the insufficiency's genetic cause.

With the help of Eleanor's parents, I sketched out a family tree, verifying that there were no close relatives known to have been diagnosed with abnormal development of the sex organs. Finally, I explained the tests that the endocrinologist and I would be ordering and what they could reveal.

"Sexual differentiation of the fetus as a male or female happens early on, and it depends on the well-timed delivery of signaling molecules known as *hormones*," I began. "Some of these hormones are secreted in the brain, and others in the adrenal glands, but all of them trigger the activation of transcription factors—molecules that control the way genes express proteins, the functional building blocks of tissue.

> "The genetic exams we'll do will start with a blood test. Examination of your blood will reveal whether your DNA codes for certain important hormones and whether the hormones produced based on these genetic instructions developed as expected. We'll also be looking for any mutations in transcription factors that may have led to gene expression that differs from the norm. Once we have this, we will be closer to having a full picture."
>
> Eleanor's father, Phillip, spoke up at this point. "How long will getting an answer take?" he asked.
>
> I responded, "We will do our best to speed things up, but usually this takes a couple months."

Once the genetic tests were completed and analyzed, Eleanor received jarring news: although she had considered herself to be female, she carried both an X and a Y chromosome. In terms of her genotype, she was male. In terms of her phenotype, she both lacked female sex organs and did not have outward signs of any male genitalia.

Management of Eleanor's DSD would take place over a period of years, with a focus on supporting healthy physical development, gender identity, and emotional well-being—while optimizing her future possibilities for sexual fulfillment and, if desired, for having children. Care of this kind relies on a multidisciplinary team—endocrinologists, surgeons, psychologists, genetic counselors, and social workers—who guide the patient and family through both the medical and emotional aspects of the journey.

Ultimately, the goal of this work, involving medical and psychological professionals, the family, and the affected patient, would be to generate an individualized and ethical approach that promotes the affected child's autonomy and future well-being. This was not the end of the story.

The team, led by the endocrinologist, together with laboratory researchers, identified the gene and confirmed through experimentation that the mutation caused sex reversal and was indeed pathogenic. In addition to ongoing medical and psychological care, Eleanor's parents would come back for additional genetic counseling, as the mutation increased the risk of another child with the disease. This dramatic story serves as a reminder that genetics is not just about science but about people—about identity, uncertainty, and the weight of decisions made in moments of crisis. In these fragile hours, our role as clinicians, counselors, and scientists must be not only to diagnose but also to hold space for the complexity and to walk with families through it.

This chapter may end here, but the connections Dr. Banne made with these families do not. They stay with him. Always.

Chapter 17
Nature and Nurture Together: The Challenge of Multifactorial Conditions

Until this point in the book, we have primarily focused on binary properties. Like Mendel's smooth or wrinkled peas or people who have cystic fibrosis, these phenotypes, driven by a single gene, are associated with just two discrete outcomes. Binary properties are easy to distinguish and easy to trace to their genetic roots.

However, there are also genetic properties that exist on a spectrum. In this chapter, we will explore a grey zone of genetics, where the definitions and distinctions between binary and continuous properties become blurred. While this may be more challenging to analyze, it reflects the real world, the complexity of genetic counseling, and the impact that counseling can have on the lives of patients.

> In my long years of clinical practice, I have learned that not every condition has a clear-cut genetic diagnosis. I have also learned that this was sometimes difficult for patients to accept. The Dorham family was a good example.
>
> Silvia and Albert Dorham arrived for their appointment right on time, accompanied by two young children. Although we had never met, it was easy for me to guess what had brought this family to the clinic. All four were redheads, but that wasn't the only trait they shared. Both the mother and the father were extremely overweight. The older of the two children—a 10-year-old boy who shyly introduced himself to me as George—showed signs of juvenile obesity. George's younger sister, on the other hand, seemed to be of normal weight for her age and height. After everyone had taken their seats, I turned to the parents and asked what had brought them to the clinic.
>
> "Our pediatrician referred us to you when George's weight began to creep up past what's normal for his age," Silvia began. "Since there's a history of weight problems in our family, she thinks that George's problem might be caused by a particular gene."
>
> Albert then chimed in. "Of course, we understand that what causes fat is eating too much or not exercising. I'm a paid-up member in that club," he said, his half-smile betraying a hint of embarrassment. "But our doctor told us that if George's weight is caused by a genetic mutation, then it's actually good thing: we'd be talking about something that can be controlled with drugs."

I understood the motivation behind the pediatrician's referral. Obesity is a major and growing health problem for adults and children, one that drives an increased risk for cardiovascular disease and diabetes. The problem is, the more common a

E. Banne, B.-Z. Shilo, *Dialogue With Our Genes*,
https://doi.org/10.1007/978-981-95-9571-6_17

condition—and unfortunately, obesity is very common—the more likely it is that an individual's weight problem is rooted in several interrelated factors, both genetic and environmental.

Not that Albert was wrong about the fact that obesity can sometimes be treated medically. Back in the 1990s, Jeffrey M. Friedman and his colleagues at the Rockefeller University discovered how leptin, a hormone secreted by fat tissue, flips a switch, a receptor that turns on our ability to feel full after eating. When a mutation in a single gene causes this receptor to not function, the person carrying the variant never feels satiated and, as a result, eats uncontrollably.

Such individuals become obese from a very early age. Because of this, I began the family's intake exam with a very specific question.

"Has George always been overweight?" I asked.

"Not at first," Silvia replied. "We started noticing some weight gain around his eighth birthday, and his weight has been going up ever since."

Taking quiet note of this information, I continued my interview, which allowed me to sketch out a three-generation family tree showing that both sets of grandparents were also significantly overweight. A questionnaire focusing on George's eating habits showed that he was not constantly hungry, did not eat inappropriately large amounts, and often left food on his plate. Taken together, these facts indicated that George's weight problem was not monogenic—that is, it was not rooted in a single gene mutation that affected the leptin pathway function. The family, however, did not want to give up hope that George's problem would be easy to solve.

"Our pediatrician told us that there's a drug for people whose genes make them overweight, and obviously this trait has been passed down through our family," Silvia argued. "So why not at least try drug treatment and see if it helps?"

I explained that it was not so simple. "George's symptoms and history don't fit the criteria for monogenic obesity testing," I said. "That means that your health insurance won't cover it, and I have to tell you, the test is very expensive."

"We'd be willing to pay privately," Albert declared. "No offense, Doctor, I'm sure you know how these things are done, but we really respect our pediatrician and want to follow her advice. I mean, my wife and I both know what it's like to grow up with a weight problem. We want to break the cycle if possible, and we want to do it sooner rather than later."

After some more discussion, I realized that the parents were adamant about pursuing the option of private testing, so I prepared the necessary paperwork. At the same time, I referred them to something that their health plan did cover: a consultation at our medical center's obesity clinic. I explained that because of the family history, their younger daughter was also at high risk for weight gain as she grew older. This touched a nerve, but it had to be said: behavior modification is probably the most important thing that the couple could do for their children's health at this time.

Silvia and Albert assured me that they would call for an appointment, but as they left my office, I could see that they were deeply disappointed.

As explained above, the term *monogenic obesity* refers to severe weight gain resulting from a mutation or deficiency of a single gene. Such mutations fall into two broad categories. First, obesity can be triggered by the mutation of individual genes that play a role in the leptin–melanocortin-4 (MC4R) receptor pathway. Second, obesity can also be triggered by single-gene mutations that affect the development of a brain region called the *hypothalamus*. In most cases, however, obesity is a complex syndrome involving multiple genes as well as environmental factors. Societal prevalence of multifactorial, as opposed to monogenetic, obesity led me to believe that when the test results returned, George's weight problem would be confirmed as a complex syndrome without a single cause.

> After waiting a long time for private monogenic testing, it took a further 2 months for Silvia and Albert to receive the results. Then, after waiting for an opening at my clinic, the family returned for a follow-up visit. The passage of time had not tempered the parents' hopes. It had even increased their desire to push for drug treatment that would "fix" their son.
>
> Handing me the test results, Albert presented me with a note upon which he had written "setmelanotide," the name of the drug used to treat monogenic obesity. Even though I knew that this medication would not be useful in George's case, I felt it important to first acknowledge the parents' strong desire to help.
>
> "I'm glad you were able to arrange for the private screening, which clarifies what we are dealing with," I began. "The test results show that George does not carry a single-gene variant that interferes with melanocortin signaling. Setmelanotide, the drug your doctor told you about, is designed to target the MC4R receptor and reactivate it when it is not functioning. In George's case, I'm sorry to say that this drug will have no clinical effect."
>
> Silvia looked visibly deflated. "So where do we go from here?"
>
> "We'll need to take a look for a wider picture, one that takes into account the activity of multiple genes as well as factors in the environment," I answered. "Luckily, genetic science has given us a number of approaches for doing just that."

The first scientist to study multifactorial inheritance was Sir Francis Galton. A cousin of the father of evolution, Charles Darwin, Galton was a contemporary of Gregor Mendel, the monk whose experiments with pea plants revealed the principles behind the inheritance of dominant and recessive traits.

Unlike Mendel's observations, which identified strict ratios by which some traits were passed down while others were not, Galton discovered that some phenotypes, such as human height, vary along a continuum. If plotted on a graph, they create a bell-shaped curve, with the very tall and very short situated at the extremes, and those of intermediate height (who make up the majority) bunched in the middle. Such traits are influenced by a wide range of genetic and environmental factors. For example, a person might not achieve the full height potential coded into their genes due to environmental factors, such as compromised nutrition during childhood. Such traits cannot be broken down into discrete "yes" or "no" categories.

Traits characterized by this type of continuous variation are called *quantitative traits*. Quantitative traits only manifest when a critical threshold of combined factors has been reached. The severity of the symptoms corresponds to the degree to which a combination of genetic and environmental factors exceeds the threshold.

Sometimes, environmental factors that contribute to this quantitative dynamic affect the onset of genetic disease. One example is cleft palate, a phenomenon in which a baby is born with an opening or split in the roof of the mouth due to tissues that did not fuse together during fetal development. Although cleft palate is rooted in genetics, environmental factors increase the risk. These include prenatal nutritional deficits—particularly a lack of folic acid in the mother's diet—and smoking during pregnancy.

Another multifactorial condition is coronary artery disease. The leading cause of mortality in the Western world, coronary artery disease, is known to run in families. Lifestyle and environmental factors, however, also contribute. Studies in the US have shown that it occurs more often in men than in women and that African Americans are at greater risk of death due to this condition than Caucasians or Asians are. Additional health conditions such as social factors, obesity, high blood pressure, and even gum disease raise the risk, making coronary artery disease a classic model of a multifactorial disorder. Type 2 diabetes, which runs in families, is another example of a multifactorial condition that cannot be traced to a single gene variation.

17.1 Gene Plus Gene Plus Gene...

So how do multiple genetic factors come together to cause disease? In multifactorial conditions, an additive interaction between mutations often causes a phenotype to occur. While one defective gene may not trigger symptoms, the influence of a second or third genetic mutation may trigger disease onset.

How multigene interactions play out is not always obvious. When researchers tried to identify the genetic variations that contribute to type 2 diabetes, for example, they reasonably enough began by investigating genes known to impact the glucose-insulin pathway, a set of molecular interactions that, in patients with this metabolic disorder, fails to maintain the proper level of sugar (glucose) in the blood.

Then in 2007, a study led by British scientists revealed the involvement of an unexpected additional gene mutation associated with type 2 diabetes onset. The scientists found a correlation between this condition and a variant of *fat mass and obesity-associated genes* (*FTO*). They discovered that 16% of adults with two mutated copies of this gene had significantly increased odds of becoming obese, which strongly correlates with type 2 diabetes. According to the statistics, this same correlation holds true for children over the age of seven. Although it is now well known that people who are overweight have a heightened risk for type 2 diabetes, this was a surprising discovery at the time.

The point here is not that a mutation of the FTO gene causes diabetes. On the contrary, it is the obesity triggered by this mutation that contributes to diabetes onset. And obesity has its own molecular pedigree; since the British study, additional gene-linked phenomena have been shown to play a role in obesity, including alterations in a transcription factor called *TCF7L2* and changes in the structure of pancreatic beta-cell proteins.

The main takeaway is this: multifactorial genetic diseases cannot be traced to a single abnormal gene. Instead, they are the product of a complex dance involving several molecular interactions and environmental factors.

17.2 Searching for the Multifactorial Origin Story

As we have seen, the combined influence of nature (genes) and nurture (the environment) can make it difficult to identify causes of multifactorial conditions. But that doesn't mean that scientists and doctors don't try.

Take celiac disease, an incurable condition that causes a damaging immune reaction in response to the gluten found in wheat, barley, and rye. Searching for celiac's cause, scientists first focused on gene variants known to play a role in autoimmune conditions, diseases in which a strong, off-target reaction of the body's defense system causes immune cells to attack and destroy healthy tissues.

Scientists discovered a strong correlation between celiac and a family of molecules called *human leukocyte antigens*, or *HLA*s. Variants in specific HLAs are extremely common, seen in 30% of the European population, which is far greater than the miniscule cohort of Europeans who are actually diagnosed with celiac disease. But all celiac patients have these variants. This begs the question: what causes some people who carry specific HLA subtypes to develop celiac while others do not?

One might assume that individuals with celiac symptoms harbor specific HLA subtypes alongside additional gene variants. Perhaps they were also exposed to environmental stresses that cause the condition to manifest. The jury is still out on what causes it. But although the multifactorial mystery of celiac onset has yet to be solved, the search has not been a total loss. By proving that the condition is linked to specific HLA subtypes, scientists provided a tool for clinical diagnosis. These same variations can serve as *biomarkers*, molecular flags that help clinicians make a positive celiac diagnosis in patients suspected of having the disorder.

But what if doctors are trying to identify a multifactorial disorder and there's no connection between the symptoms and any known gene variant? In such cases, advanced technologies developed in research labs can come to the rescue.

As described earlier in this book, it is possible to track tiny changes that occur in DNA. These changes, known as *single nucleotide polymorphisms*, or *SNP*s (pronounced "snips"), are the most common type of genetic variation in humans. Each SNP represents a difference in a single DNA building block, called a *nucleotide*. For example, a SNP may replace the nucleotide cytosine (C) with the nucleotide thymine (T) in a certain stretch of DNA.

Most SNPs have no effect on health or development. Some of these variants, however, have proven to be clinically important—we mentioned SNC's (C is for change) in a previous chapter. SNPs can help predict an individual's response to certain drugs, susceptibility to environmental toxins, and risk of developing diseases.

SNPs are also helpful in the search for the subtle variations that drive multifactorial genetic conditions. This is because they can be compared across huge population groups, thereby making it possible to identify individuals who carry a similar variant and display comparable symptoms.

17.3 The Wider Genomic Perspective

By its very nature, SNP data reflects the involvement of a huge number of genes that may appear in a normal or mutated form. To mine this wealth of data and get closer to a diagnosis for multifactorial disease, scientists developed an approach known as *genome-wide association studies*, or *GWAS*.

Genome-wide association studies have been used to identify variants associated with conditions including diabetes, heart disease, Parkinson's disease, schizophrenia, and Crohn's disease. GWAS is a population-wide approach. Therefore, researchers can use it to help predict which prevention and treatment strategies will be effective for certain groups.

GWAS is also a remarkably powerful tool for clarifying the genetic basis of multifactorial conditions. That is exactly what was discussed the next time the Dorham family visited the clinic.

Silvia, Albert, and George returned to my office for a routine checkup. The parents now understood that the hereditary obesity in their family could not be attributed to a single mutation that could be treated with a single drug. But they still hoped that the wider genetic picture would give them more than just a clinical diagnosis. They wanted me to provide them with the winning strategy for combating their son's weight issues.

"Genetic testing can provide us with a lot of important information, but it can't tell us everything we want to know about every patient," I said. "This is particularly true for a condition like obesity, which is extremely common and can be triggered by many different factors."

I went on to relate how, together with a friend who did genetic research at the university, I had discovered something interesting using GWAS.

"Using a database that contains findings from genetic studies in labs and clinics all over the world, my colleague and I were able to identify variants connected to obesity. Not surprisingly, one was a mutation in a gene that disrupts the melanocortin pathway, the same variant that eliminates our ability to feel satiated and triggers obesity from a young age. The mutated gene was *proopiomelanocortin*, or *POMC*. For some time, it has been known that POMC causes severe monogenic obesity, the drastic weight gain associated with a single gene variation. But in our research, we determined that this gene has relevance for your family, even if George's obesity is multifactorial and cannot be linked to this single genetic variation."

The parents looked at me quizzically.

"It turns out that the same area of the genome that carries single-nucleotide variations affecting the melanocortin pathway is also associated with red hair," I continued. "While it's difficult to say exactly which combination of genetic interactions led to your family's phenotype, most patients who carry two copies of a POMC mutation, one from each parent, are not only obese. They are also redheads."

Silvia and Albert looked at each other, then at their red-haired son, and grinned. It seemed that this scientific information added some much-needed levity to a heavy discussion.

Albert then turned to me and asked, "Can the information in this database help us predict the future about other children? I mean, maybe our daughter doesn't have all the genetic mutations that would cause her to develop a serious weight issue."

By analyzing hundreds or thousands of SNP variations across a person's genome, it is indeed possible to determine how many SNPs exist that are associated with a particular condition. However, I explained to the couple why this approach, called a *polygenic risk score*, cannot yet provide a diagnosis. It cannot definitively rule out the risk for a disease, making it a poor indicator for making treatment decisions.

"I believe there will be technological progress in the future," I said. "But in the meantime, your best bet is the old-fashioned approach—dietary adjustment, regular exercise, and medications for those who can receive them, taken under professional supervision. Nowadays, there are several medications that can be used to treat obesity in adults, including GLP-1 agonists. But as of now, there are limited options for treating obesity in children, to be used with caution."

We went on to discuss various approaches for managing obesity over the long term, including surgical intervention. I referred them to a clinical specialist in the field. Still, I sensed that Silvia and Albert were holding on to the hope of discovering a single gene that could be targeted with medicine.

I wish I could have offered the Dorham family a quick fix. Instead, the parents left my office armed with the realization that no matter how many factors contributed to their obesity, it was up to them to make the adjustments needed to ensure the best health outcomes for their children.

Chapter 18
When Genetic Diagnosis Is a Mid-Life Crisis: The Special Case of Late-Onset Disorders

When Becca arrived at the clinic, she gave the impression of a woman used to being in control. But when it was time for her appointment, Becca's difficulty rising from her chair and her heavy gait as she stepped into my office betrayed the lack of physical control that had brought her that day.

Settling into her seat, Becca handed me an organized file containing her doctor's referral and a complete medical history. She explained how as manager of a global team of engineers and the divorced mother of two young children, she was no stranger to fatigue. But a couple of years ago—shortly after her divorce—she began to notice an increasing feeling of heaviness in her legs.

"At first, I dismissed it as stress, or the result of letting my gym membership lapse after the divorce," she said. "But then I found myself stumbling in ways I couldn't explain away, with my left foot seeming to drag me down. Finally, last month, when my left foot completely refused to cooperate, I went to my family doctor, who sent me to a neurologist. He did some tests and told me I'm experiencing something called *drop foot*. He said that it has a genetic cause that I might have already passed down to my children."

Although her tone was direct and businesslike, I handed Becca a box of tissues as she wiped away a tear.

My experience as a genetic counselor has taught me that when a patient walks in with that level of fear, every word I say could be taken as a confirmation of their darkest suspicions. From her tense posture and clenched hands, I could tell that Becca had already braced herself for the worst.

"Let's talk first," I began, keeping my voice gentle. "Before we go into genetics or diagnosis, please tell me more about yourself."

Becca exhaled slightly, as if appreciating the pause. She nodded, and we started from the beginning. I asked about her family, her job, her children—pieces of her life that didn't make it into the sterile referral note. With each question, her story slowly unfolded. Becca was of Jewish Persian origin. Her siblings, both younger than her, were married with children, and neither had any health concerns. Her father had passed away tragically in a car accident when he was 60, but her mother was still alive and well. No one in her family had a history of neurological problems.

As she spoke, I could see her tension ease a little, though her eyes still flickered with uncertainty. There was no obvious family pattern to explain her condition. Nothing in her answers suggested an inherited disorder, but I knew we couldn't rule anything out yet.

I moved on to the physical examination, focusing on her feet, gently testing their strength, and observing their movement. It didn't take long to spot what had been troubling her for months—her left foot, stiff and uncooperative, showing the classic signs of foot drop. As I continued, I noted mild weakness in her other muscles. But the quadriceps? They were unaffected. That small detail, the sparing of the quadriceps, was a bright arrow pointing toward a diagnosis.

E. Banne, B.-Z. Shilo, *Dialogue With Our Genes*,
https://doi.org/10.1007/978-981-95-9571-6_18

I paused, leaning back slightly. "You're experiencing some muscle weakness, but your quadriceps are strong. That's interesting," I said, more to myself than to her. "There's a condition that fits this pattern."

Becca's eyes widened slightly. She was holding her breath, waiting for the next words to leave my mouth.

"I'd like to suggest a genetic test," I continued, meeting her gaze. "There's a condition called *GNE myopathy* that can present like this, particularly in people of Persian descent. A specific variant of this gene is more common in people with your background. That your quadriceps aren't affected makes this a real possibility."

It took a month for the results to come back. When the call came from the lab confirming my suspicions, I knew that for Becca, the news would be life-altering.

When Becca arrived for her follow-up, the same anxiety was etched into her face, but this time her eyes held a quiet determination.

"Becca," I said, carefully choosing my words. "The test results came back. You do have a genetic condition—GNE myopathy."

I explained the condition as simply as I could, the impact it might have on her muscles, and the long-term outlook. I told her that while GNE myopathy wasn't curable, there were ways to manage it. I watched as Becca absorbed it all, and I saw something change in her. The unknown was no longer looming over her. She knew what she was facing, and while it wasn't what she had hoped for, it was something she could begin to understand and accept.

In this chapter, we will use GNE myopathy as a doorway through which to discuss the broader challenges of diagnosing late-onset genetic disorders. Before diving into the complexities of GNE myopathy specifically, let's define the scope of the topic. *Late-onset genetic disorders* are conditions that do not manifest during childhood, typically revealing their symptoms only after the age of 18 and, at times, only during middle age. In their classic form, these disorders remain under the radar for much of an individual's life, allowing affected people to lead seemingly normal lives until the first signs appear.

For many individuals, the first decades of life are entirely normal. They have no reason to believe they are not healthy and may go on to marry, have children, and live fully functioning lives. Because these conditions often appear so late, affected individuals are typically unaware of the genetic disorder they carry, and neither they nor their doctors have any early indication of the illness.

Even once symptoms begin to appear, these conditions are commonly mistaken for multifactorial disorders, meaning they seem to result from a combination of genetic and environmental factors. For example, a person may experience mild muscle weakness or cognitive decline, which can easily be attributed to natural aging, lack of physical activity, or lifestyle. This overlap of symptoms with common age-related conditions makes it incredibly difficult for doctors to suspect a genetic cause in the early stages.

A subtle muscle weakness might be dismissed as the result of inactivity, or early cognitive issues could be chalked up to stress or age. Without considering the possibility of a single-gene disorder, clinicians—usually neurologists—may not consider referring these patients for genetic counseling or testing. This can lead to years of misdiagnosis or no diagnosis at all.

The key challenge, therefore, is that symptoms in late-onset disorders can resemble those seen in the general population due to aging or common health issues. For instance, in early-onset Alzheimer's disease, which may begin before age 60, the first person in a family to show symptoms is often a grandparent, as the condition usually emerges in later adulthood. While dementia is often associated with older adults, early-onset cases can be easily disregarded as unusual but not concerning. Similar challenges arise in conditions like Huntington's disease, Parkinson's disease, and certain genetic cancers where early symptoms don't immediately suggest a genetic disorder.

One important clue that can guide clinicians is the earlier-than-expected onset of symptoms. While late-onset genetic disorders typically appear in adulthood, their genetic basis may cause them to manifest slightly earlier than similar nongenetic conditions. For example, GNE myopathy might cause muscle weakness earlier than one would expect based solely on age-related deterioration. Recognizing this discrepancy—when a patient presents with symptoms earlier than would be typical—can alert clinicians to the possibility of an underlying genetic cause.

Tracing inheritance patterns in these disorders presents yet another challenge. Geneticists rely heavily on family histories, but family trees for late-onset disorders are often incomplete or misleading. Many individuals who carried the mutation may have passed away before showing symptoms or being diagnosed, especially in dominant disorders where one parent may have died before receiving a diagnosis. In the case of recessive inheritance, there may be no obvious family history of the disease, as carriers remain unaffected, leading to further diagnostic hurdles.

Moreover, *variable penetrance*, or the proportion of individuals carrying a particular variant (or allele) of a gene that also expresses an associated trait, and *expressivity*, or the degree to which a trait is seen in individuals, add another layer of complexity. Even within the same family, individuals carrying the same genetic mutation may show a wide range of symptoms or, in some cases, no symptoms at all. This variability makes it difficult to draw clear conclusions from family trees, as the disorder may express itself in unpredictable ways.

A delayed or missed diagnosis doesn't just affect the patient; it has implications for their relatives as well. Once a genetic disorder is identified, it becomes clear that family members may also be at risk, especially siblings or children who may carry the same mutation but remain asymptomatic. Early diagnosis can be important not only for managing the affected individual's health but also for offering genetic counseling and testing to at-risk family members, potentially changing the course of their health management.

18.1 The Impact of a Late-Onset Diagnosis

Becca's syndrome, GNE myopathy, is a rare genetic disorder that primarily affects the muscles we use for movement. While it often appears in early adulthood, those with the condition experience a slow but progressive muscle weakness that worsens over time. Many patients first notice a problem when they experience difficulty lifting the front part of the foot due to muscle weakness in the lower leg. Over time, this weakness can spread to the upper legs, shoulders, and hands, making everyday activities increasingly difficult.

The cause of GNE myopathy lies in a gene called *GNE*, which provides instructions for an enzyme responsible for producing sialic acid, a crucial sugar molecule. Sialic acid helps cells communicate, move, and stick to one another, functions that are essential for maintaining healthy muscle tissue. In people with GNE myopathy, a mutation disrupts normal production of sialic acid, leading to muscle degeneration.

While GNE myopathy is a debilitating disease, it is also very rare, affecting fewer than nine in one million people worldwide. The condition is more common in certain populations, such as Japanese, Iranian Jewish, and people from the Indian subcontinent, but many cases remain undiagnosed due to its slow progression and the fact that it doesn't appear until adulthood.

Patients often describe the diagnosis of GNE myopathy as a moment that alters their understanding of themselves and their future. Many are in the prime of their lives when symptoms first appear, which can range from a slight limp to difficulty running or even walking.

Some patients tell us that their first reaction is disbelief; after all, they were active, healthy individuals, some of whom participated in sports and physical activities. As the disease progresses, they begin to gradually lose mobility, a process that forces them to adjust both their physical capabilities and their expectations for the future.

18.2 The Challenge of Treatment

Because GNE myopathy is caused by a genetic mutation, researchers are focusing on therapies that can correct the underlying genetic problem or increase the body's production of sialic acid. One promising approach being pursued in science labs is *gene therapy*, where researchers deliver a healthy copy of the GNE gene to muscle cells in the hope of restoring their ability to produce sialic acid. Early experiments in mice have shown some success; treated animals exhibit improved muscle function and strength.

In addition to gene therapy, researchers are also exploring sialic acid supplementation as a potential treatment. In studies on mice, feeding animals compounds that increase sialic acid levels has helped slow the progression of muscle degeneration, offering hope that a similar approach could work in humans.

However, there is still much to learn. One of the biggest challenges is developing reliable animal models that accurately mimic the disease as it occurs in humans. Current models don't always show the full range of symptoms seen in patients, making it difficult to test treatments effectively.

Despite these hurdles, researchers remain optimistic that with continued study, new treatments for GNE myopathy will emerge. While there is no cure yet, advancements in gene therapy and sialic acid research may one day provide relief for those living with this rare condition.

18.3 Living with GNE Myopathy: Resilience in the Face of Decline

Living with GNE myopathy often requires individuals to make gradual yet significant adjustments to their lives. Early on, patients may try as much as possible to continue with their usual activities—working, raising families, and maintaining hobbies—despite their growing physical limitations. But over time, many come to rely on aids such as leg braces, canes, and electric scooters. While some lose the ability to run or walk long distances, they find new ways to navigate life with creativity and resilience.

For some, the emotional journey of accepting their condition is just as challenging as the physical symptoms. GNE myopathy's slow progression means that patients must confront their changing abilities and, in some cases, redefine what they consider strength and endurance. Patients may initially feel overwhelmed by their diagnosis, then gradually come to terms with it and understand that resilience isn't about pushing past the pain or denying the reality of their condition; it's about accepting it and continuing to live fully within new limitations.

Maintaining optimism and a positive outlook can play a crucial role in how patients cope with the disease. Many describe finding joy in the small victories, whether it's completing a walk with assistance or continuing to pursue hobbies like cooking or traveling within their limits. These moments, patients say, offer a sense of accomplishment that fuels their ability to keep moving forward, even as the disease progresses.

Genetic testing is a valuable tool for understanding the root causes of diseases, even those that appear later in life. For patients already showing symptoms, confirming the genetic basis of their condition can provide clarity, guide treatment, and offer a more accurate prognosis. However, testing for late-onset genetic disorders raises significant ethical questions, especially when it comes to testing children and managing the information obtained from adult testing.

18.4 Late Onset and the Ethics of Pediatric Genetic Testing

Late-onset genetic disorders present a unique ethical dilemma, particularly in pediatric testing. Guidelines from medical organizations like the American Academy of Pediatrics recommend against testing children for conditions that will not appear until adulthood, and with good reason. Testing children for late-onset conditions such as GNE myopathy removes their ability to decide for themselves as adults. While parents may want answers, children should ultimately have the right to make their own decision when they are mature enough to understand the implications. This respects their future autonomy and allows them to make an informed choice when they reach adulthood.

For most late-onset disorders, there is no preventive treatment available in childhood. Therefore, knowing that a child is at risk provides no immediate health benefit but may impose a psychological burden. Without actionable interventions, the potential harm of testing—such as anxiety or altered self-perception—outweighs any benefit. Furthermore, children are not capable of fully understanding the complex and long-term implications of genetic testing. Even if the condition is explained to them, they are unlikely to grasp its future significance. As such, true informed consent cannot be obtained from children, further supporting the need to delay testing until adulthood.

When it comes to adults, predictive genetic testing for late-onset disorders offers insights but also introduces new ethical complexities.

Unlike children, adults can provide informed consent, but they may not always fully comprehend the long-term psychological and lifestyle implications of testing. Some individuals who receive a positive test result experience regret or report they hadn't anticipated the emotional impact. Anxiety, depression, and an altered sense of life planning are all potential responses. For instance, a young woman who learns she will develop GNE myopathy might feel discouraged from pursuing a long-term dream, like becoming a doctor, fearing she won't have time to fulfill it.

Privacy remains a critical concern, as genetic information recorded in medical files raises concerns about potential access by insurance companies or employers. This may result in discrimination against completely healthy individuals solely based on their genetic predisposition. Imagine a talented professional on the path to a high managerial position being passed over due to concerns about future health issues. In the wrong hands, genetic data can lead to significant social and financial harm.

Predictive testing can complicate relationships. In some cases, a person may wish to know their genetic status, while their partner or family members would prefer to remain unaware. Decisions about whether to share test results or consider genetic information in family planning add additional layers of complexity. Should genetic status guide reproductive decisions? These are deeply personal questions that each individual or couple must address, often with the support of genetic counselors.

While genetic testing offers invaluable insight into late-onset disorders, it also raises profound ethical challenges. Testing children is generally discouraged to protect their autonomy, while adults must consider the potential for discrimination, psychological effects, and the complexities that come with sharing genetic information. As our understanding of genetic testing's benefits and drawbacks grows, so does our responsibility to approach it with care, sensitivity, and respect for individual choice.

Chapter 19
The Genetics of Inherited Cancer

Cancer is a disease in which some of the body's cells grow uncontrollably, crowding out normal cells and sometimes spreading to other parts of the body. Cancer makes it hard for your body to work the way it should. Whatever form it takes, cancer is a serious diagnosis.

Cancer, which can appear in many different tissues and can progress along different trajectories, emerges in response to both genetic and epigenetic factors. Most cancers are caused by changes in genes that control the way cells grow and multiply. Random mistakes can appear in the DNA passed down to daughter cells during cellular replication. Environmental factors like the chemicals in tobacco or exposure to the sun's UV rays can also trigger genetic changes. Aging is highly correlated with an uptick in cancer risk, indicating that cancer can be caused by the cumulative effects of genetic and environmental factors over time.

Two distinct categories of genes are involved in cancer onset and development. Under normal conditions, the job of one of these categories—the *oncogenes*—is to regulate the controlled division of cells. In their mutated state, however, oncogenes become overly active or are expressed at inappropriate times or locations. This drives pathological unrestrained cell division. Oncogenes are dominant and typically lead to cancer when activated. But since these mutations occur in body cells rather than in reproductive egg or sperm, oncogene mutations are not inherited; they will not be passed down to offspring. As such, we will not be addressing oncogene mutations in the context of this book.

The other type of cancer-causing genes is known as *tumor suppressors*. Tumor suppressor mutations can be inherited, and in 5–10% of cases, cancer is caused by tumor suppressor mutations passed down from parent to child. In this chapter, we will be looking at what tumor suppressors are, what happens when they become mutated, and what implications they have for genetic counseling as well as clinical cancer treatment.

E. Banne, B.-Z. Shilo, *Dialogue With Our Genes*,
https://doi.org/10.1007/978-981-95-9571-6_19

19.1 What Are Tumor Suppressor Genes?

Tumor suppressor genes are cells' first line of defense against cancer onset. In their normal, nonmutated form, they produce proteins whose function is to identify changes to DNA that can drive a healthy cell toward malignancy. Once DNA damage is identified, tumor suppressor proteins have two modes of protective action: they either repair the damage or they trigger cell death, which prevents dangerous cancer cells from replicating.

Some tumor suppressor genes produce proteins that help maintain the stability of DNA, preventing mutations from building up in other genes that can promote cancer. This proactive protection limits the production of abnormal proteins that drive tumor formation.

With such an important role to play, when tumor suppressors become mutated, the results can be disastrous. Just one mutation, in the version of a gene inherited from either the mother or the father, can initiate a process that may eventually lead to cancer in an otherwise normal cell.

Luckily, most mutations in tumor suppressor genes behave in a recessive manner at the cellular level, which provides a degree of protection. Individuals who inherit one altered copy of a tumor suppressor gene still have a second, functioning copy that can maintain normal cell regulation. However, cancer can develop if the remaining healthy copy becomes damaged over time—a process often referred to as a "second hit." When both copies of the gene are no longer functioning within a cell, its ability to control growth and repair DNA is compromised, increasing the risk that the cell will become cancerous.

Genetic testing for mutations in tumor suppressor genes can contribute to earlier cancer diagnosis and improved prevention. For example, people who inherit a mutation in one copy of a gene such as MLH1 may have Lynch syndrome, which increases the risk of colorectal, uterine, stomach, urinary tract, and other cancers. Identifying such a mutation before symptoms appear can prompt doctors to begin screening early, improving the chances of detecting cancer at a treatable stage. A rarer and distinct condition, called constitutional mismatch repair deficiency (CMMRD), occurs when both copies of the gene are altered and usually presents in childhood with multiple cancers.

Becoming aware of a genetic predisposition for cancer is not just important for the individual patient. As the story of one of Dr. Banne's patients illustrates, genetic testing can provide lifesaving information for the next generation as well.

> "How can I help you today?" I asked Angela, a 45-year-old woman who arrived at my clinic. Like many of my first-time patients, she seemed worried. I waited for her to relate her story, which, despite its happy ending, was indeed worrisome.
>
> "I came to see you because I almost died of ovarian cancer," she began. "In fact, I would never have been able to be here if my gynecologist hadn't been suspicious about my family's cancer history. Every one of my relatives who got cancer refused the genetic tests that might have identified their tumors' cause. For the sake of my children, I don't want to make the same mistake."

She described the breast cancers diagnosed in an 80-year-old aunt and, more unusually, in Angela's 60-year-old brother. She added that a cousin on her mother's side was undergoing treatment for metastatic ovarian cancer. Like her aunt and brother, this cousin declined recommendations for genetic testing, saying she was too weak for "unnecessary" procedures.

"My gynecologist told me that given this family history, I need to be vigilant about a potential cancer risk," Angela continued. "She noted a very slight change in my menstrual cycle a couple of months ago and immediately sent me for an ultrasound, which showed something strange in my left ovary. It turned out to be early-stage cancer. A day later, I was on the operating table having the ovary removed. It literally saved my life."

Angela was remarkably lucky. Due to the nonspecific nature of early symptoms, most cases of ovarian cancer are identified at an advanced or even untreatable stage. In the pathology report Angela brought to our meeting, the removed ovary was classified as "solid, pseudo-endometrioid, and transitional cell carcinoma-like morphology (SET features)," a diagnosis typical of ovarian cancer with a genetic cause. Given her family history, this was unsurprising. But a question remained: what made Angela decide to come to the genetic counseling clinic now?

"The fact is, Doctor, I have a daughter and a son," she said. "I want to give both my children the information they need to protect themselves. Even if genetic testing can't prevent cancer, it may give them what I got just by chance: the earliest possible treatment."

As I nodded and began to assemble the necessary paperwork, I thought to myself, *if only more people were aware of how useful, and even lifesaving, timely genetic testing can be.*

In some disease conditions, specific genetic mutations cause the disease to manifest 100% of the time. With cancer, however, the connection between genetic mutation and disease onset is a matter of probability. This means that a patient identified as carrying a cancer-associated genetic allele may not get cancer but has a higher risk of doing so.

Testing revealed that Angela carried an inherited defect in the BRCA1 gene. In its normal, nonmutated form, BRCA1 is a tumor suppressor. It carries instructions for making a protein that prevents cells from growing and dividing too rapidly or in an uncontrolled way. BRCA1 proteins help repair damaged DNA. But when the BRCA1 gene is mutated, its protective function is lost.

Hereditary mutations like BRCA1 are passed down from parents to children and occur in approximately 1 of every 300 individuals in some populations. This makes them much more common than syndromes rooted in the random genetic errors that can occur during cellular replication or in response to environmental factors like cigarette smoking. Despite their frequency, however, familial genetic mutations most often remain hidden, as family members carrying the cancer-associated allele usually have no symptoms that would allow physicians to identify disease risk. This is what makes genetic testing so important.

Mutations in tumor suppressor genes drive the onset of cancer in very specific parts of the body. BRCA1 and BRCA2 mutations are associated with hereditary breast and ovarian cancer. If both copies of a BRCA gene are mutated, the outcome is Fanconi anemia—a severe childhood disorder distinct from the adult-onset cancers seen in carriers of a single BRCA mutation. In this juvenile genetic syndrome, bi-allelic mutations cause bone marrow failure, short stature, and malformed thumbs, as well as a heightened risk for the blood cancer known as *acute myeloid leukemia* (*AML*) and tumors in the head, neck, skin, gastrointestinal system, and genitals.

Mutations in another family of genes—MLH1, MSH2, and others—cause Lynch syndrome, a disease that increases risk for cancer onset in various tissues, including the colon, brain, gallbladder, and liver. Lynch syndrome also causes cancers to manifest at a younger age. In cases where these mutations are identified through genetic screening, patients can be monitored so that if cancer does emerge, it can be treated as soon as possible.

While mutated tumor suppressor genes do not necessarily cause cancer, they do increase the risk that a patient will develop cancer sometime in the future. In some cases, this risk increase can be quite significant. Mutations to high-risk alleles such as BRCA1 and BRCA2 can increase a woman's chances of developing cancer by up to 40–80%.

The availability of genetic tests that reveal this predisposition for cancer has created a new clinical reality. Patients who are at high risk for cancer onset don't have to wait for cancer to appear. Instead, they can manage their cancer risk and significantly reduce it.

Such patients may choose to remove as much of the at-risk tissue as possible through prophylactic surgery. Women identified with BRCA1 and BRCA2 mutations may choose to have both breasts removed in a bilateral risk-reducing mastectomy. Due to improvements in MRI surveillance technology, some clinics may not recommend this surgery, considering routine MRI follow-up to be sufficient in certain cases. Another prophylactic procedure, especially after childbearing age, is the removal of the ovaries and fallopian tubes in bilateral risk-reducing salpingo-oophorectomy, which can help reduce the risk of ovarian cancer and, possibly, breast cancer.

Unlike the clinical breast exams most women do on a regular basis, routine screening in low-risk asymptomatic women for ovarian cancer is limited. Screening procedures for ovarian cancer, which include transvaginal ultrasound (TVUS) blood tests, can generate false positives that cause women to remove their ovaries even when it isn't necessary to do so. As a result, for some women who are low-risk and lack symptoms, these screening tests may do more harm than good.

Identification of a BRCA1 mutation, however, indicates that the patient is at high risk for full-blown ovarian cancer. In such cases, prophylactic surgery—removal of the ovaries—is often recommended. Removing the fallopian tubes but keeping the ovaries has been found to reduce the risk of ovarian cancer in the general population and is being studied as an option for premenopausal women. Whatever the doctor recommends and the patient ultimately decides, a diagnosis of a BRCA1 mutation calls for mental preparation, as well as reproductive plans.

Every type of surgery involves health risks. But prophylactic surgeries for cancer prevention are challenging in a different way. Such procedures are irreversible, may generate early menopause, and may trigger trauma related to body image, motherhood, and fertility.

It is also important to know that risk-reducing surgery does not guarantee that cancer will not develop. This is because these procedures cannot remove all the tissue that may eventually become cancerous.

In a sense, Angela was in an enviable position. She was the mother of two healthy children. She did not need to make a difficult decision about prophylactic surgery, as the lifesaving surgery had already been performed. But she still had questions.

"I know that my ovarian tumor was identified and removed in time, but my body still carries the genetic mutation that caused it," Angela said. "What does that mean for managing my health in the future?"

"*Managing* is the right word," I responded. "People with cancer predisposition syndromes need to strike a balance between living their lives and not taking unnecessary risks. What this means is that there are environmental factors that may influence whether cancer will manifest or whether, in a case like yours, it will recur. We certainly don't advise patients with an inherited predisposition for cancer to smoke, or to drink as much as they want, or to spend time at the beach without putting on sunscreen. Most importantly, at-risk patients need to stay in touch with their doctors and undergo regular cancer screening."

"What exactly do I need to do?" she asked.

I laid out my recommendations, which included conducting a breast MRI and screening her second ovary. "Depending on the results," I told her, "it may be recommended that you undergo additional surgery. The purpose of this surgery would be to remove tissues where cancer may be hiding now or may appear in the future."

Angela seemed unable to breathe for a moment. I paused to let her collect herself, then continued in a gentle voice.

"Let's get you scheduled for a follow-up appointment next week, and please come with your husband," I said. "There are some important decisions to be made, but thanks to your gynecologist's excellent work, we have time to think together about what's best for you."

Picking up her purse, Angela seemed ready to end our session. But then, she turned to me. "What about my children?" she asked. "Can they be tested for the mutation now?"

Recognizing that she was expressing fear for her children's future health, I kept my voice low when I explained that even if one of the parents has a disease-associated mutation, genetic testing is generally not recommended before the age of 20, depending on the syndrome.

"Knowing about a possible cancer risk at a young age will not benefit your children in any way and may actually cause harm," I said, adding that there is no defined surveillance regimen for children and, in adulthood, such a program would be different for sons and daughters. "Test results can cause anxiety and lead them to make unhealthy decisions. Positive results that show a risk for cancer can change the course of a child's life based on misconceptions. They can inspire risky behavior or even suicidal ideation. Genetic testing done too early can turn a perfectly normal life—a life in which the disease may never appear—into a shaky life lived in the shadow of fear."

As I accompanied Angela to the office where she could schedule her next appointment, I thought about the road ahead for both her and her children. As I had told her in the session, I do not advise patients younger than 25 to get tested. What I did not tell her was that patients over the age of 18 can overrule my recommendation and choose to get tested anyway. We genetic counselors are not gatekeepers, and we accept adults' decisions.

I knew that if Angela's children eventually decided to undergo testing, they would be choosing to embark upon a complex emotional experience. Getting tested can cause anxiety. A negative result in one child can engender feelings of guilt and shame if a sibling is revealed to be a carrier. Psychological assistance is necessary in some cases. This is something that we help arrange through the genetic counseling clinic.

But there are positive responses as well. I have had patients who discovered they carried a predisposition for cancer and then decided to plan a trip around the world earlier than they would have taken otherwise. They decided to concentrate on life rather than on their fear of possible cancer-related death. As difficult as these situations are, it is a source of satisfaction for me that, in some small way, the genetic counseling I offer can put people on the path toward living their best life.

I stayed in touch with Angela and her family. It was a bumpy ride at the beginning, including removal of her second ovary and a difficult adjustment to the hormonal changes the additional surgery caused. Ultimately, however, she made peace with her decision to be tested. She remains disciplined about keeping up her personalized regimen of cancer surveillance.

Angela is aware that each of her children has a 50% chance of inheriting her pathogenic BRCA mutation. She also understands that testing cannot change the genetic inheritance with which her children were born. But she hopes that, when the time is right, they will choose to be tested. Ideally, she wants them to be free from the fear that they, too, may one day develop the cancer she faced. At the same time, she hopes they can receive a gift made possible by science: genetic self-knowledge that empowers those who carry pathogenic variants in cancer-related genes to live as healthy a life as possible.

Chapter 20
Genetic Knowledge, Ethical Dilemmas

Every prospective parent is curious about what the future holds for their unborn child, but in cases where a genetic syndrome is suspected, this curiosity may be accompanied by apprehension and fear. Modern genetic testing makes it possible to answer critical questions about fetal health. Access to this knowledge, however, raises significant questions in the realm of ethics.

The first question that needs to be answered is: Why? Why do parents want to know more about a prospective pregnancy or a pregnancy already in progress? It may seem obvious that the goal of testing is to preemptively identify a condition incompatible with life. But this definition raises questions of its own. What does *incompatible* mean? Does it refer simply to viability, that is, the expectation of a live birth? To what extent should the projected life span of the unborn child be taken into consideration? How should the parents and counselor consider issues related to the child's expected quality of life? Should the parents' quality of life—the difficulties parents will face when caring for a child with a serious illness—be part of the calculus? What counts as 'incompatible with life' is not always medically clear and often depends on values as much as biology.

Couples undergoing genetic counseling face ethical questions related to the testing process itself. Amniocentesis, for example, can provide important genetic information about an unborn child, but this knowledge comes with an increased risk of pregnancy loss in those rare cases where the testing procedure triggers a miscarriage. What kind of knowledge is worth this risk?

In the most general sense, genetic testing reveals abnormalities that can lead may reveal a wide range of developmental challenges. We can focus on two problematic developmental outcomes: intellectual disability or physical malformation. But in both categories, the severity of the symptoms that will manifest in the child at birth or will develop afterward is unclear.

Let's take an example of a physical malformation that can be identified during prenatal evaluation.

E. Banne, B.-Z. Shilo, *Dialogue With Our Genes*,
https://doi.org/10.1007/978-981-95-9571-6_20

Hydronephrosis is enlargement, or dilation, of the renal pelvis, the basin in the central part of the kidney where urine collects. Occurring in one or both kidneys, severe hydronephrosis is a major concern, because the enlargement can cause a blockage that makes it impossible for the developing fetus to produce urine. But if the condition is mild, no intervention is required; before the child is born, fetal hydronephrosis usually resolves over time on its own.

Ethical considerations become even more dramatic in the case of genetic tests that reveal only a somewhat increased risk that a child will be born with an intellectual disability. For example, a condition called *22q11.2 duplication syndrome*, in which the affected individual carries an extra copy of a small piece of chromosome 22, results in symptoms that vary widely, even among members of the same family. Children with this genetic abnormality may have slow growth leading to short stature, weak muscle tone, and, in about 10% of cases, developmental delay and intellectual disability. However, many people with the duplication syndrome have no apparent physical or intellectual problems.

When parents discover that they carry this genetic abnormality, it can cause a great deal of anxiety. In a case where the couple is not yet pregnant, they will want to avoid passing this genetic inheritance to any future children. They may decide to initiate a pregnancy through IVF so that the genetic profile of the fertilized egg can be verified prior to implantation in the womb. If the abnormality is identified after pregnancy has already begun, the ethical dilemma becomes even more fraught. In such cases, parents may choose to terminate the pregnancy.

One of the challenges associated with modern genetic testing hinges on the diversity of the developmental results that are associated with the same genetic defect. Individuals who share the same genotype (genetic profile) may have a different phenotype (the person's observable traits, from eye color and blood type to the presence of diseases and developmental abnormalities). And while specific mutations may indeed generate disease, the severity of the associated symptoms spans a spectrum of phenotypes, from minor to major challenges. This fundamental uncertainty makes any decision about whether to continue with the pregnancy or to terminate it very difficult.

Another ethical consideration is more philosophical. When parents learn that their unborn child has a genetic mutation, they must contend with the uncomfortable reality that their own genetic contribution plays a role. While the parents' symptoms might have been mild or nonexistent, they must still live with the knowledge that they passed down a mutation that puts their child at risk. If they then choose to abort that child, is that akin to aborting themselves? Would the world have been a better place if the parents had been denied the possibility of being born?

As we can see, the knowledge gained through genetic testing creates a range of highly charged questions. And answers are hard to find; although testing offers parents the power to choose how and whether to have children, or whether to continue or terminate a pregnancy, it cannot predict the future or ensure that any life-and-death decision the parents make is the right one.

20.1 Playing God?

Genetic testing can put parents into a position to play God and decide the fate of their unborn children. This begs another critical question. Should genetic counselors *withhold* some test results to encourage ethical limits to the range of decisions parents might make?

In one sense, the answer is already a resounding no. Labs that perform chromosomal microarray analysis (CMA) testing are legally required to provide parents with the results of the tests they do. It is generally agreed that parents have the right to know if testing has revealed a disease-causing mutation so that they can choose their own path forward.

That being said, there is an active discussion within the genetic counseling community about how test results should be shared. Supporting parents' right to know, while at the same time avoiding a counselor's undue influence over parental decision-making, conventional clinical guidelines call for *nondirective counseling*.

Under this general approach, which is not inscribed in any legal statutes but is followed by most responsible genetic counselors, the disclosure of genetic risk information is limited to what is necessary for the clients to make an informed decision, without direct or indirect advice. The division is clear: the counselor's job is to provide information, and it is up to the couple to decide what's best for their family.

But is it really possible for a professional geneticist to be entirely nondirective in counseling? The situations counselors face are complex, and each counselor is also a human being, influenced by their own cultural background and moral convictions.

For example, can a counselor who is a practicing Roman Catholic be completely value-neutral when discussing abortion? How can a counselor or physician with extensive experience treating children with genetic syndromes—children who are dearly loved by their parents and many of whom have grown to be healthy adults—react in a nondirective manner to parents who decide to terminate a pregnancy that might produce a child with significant medical challenges?

Let's look at one example of how this nondirective approach might play out.

> I remember the day I met Risa and Edward for the first time. This young couple had been married for 2 years and had never been pregnant. In the first moments of the consultation, Risa explained why they had been afraid to even try.
>
> "I was diagnosed with neurofibromatosis type 1 when I was a child," Risa began, "and I know this condition has a genetic cause. We're afraid of passing on the disease."
>
> Neurofibromatosis is indeed a serious condition, involving changes in skin coloring (pigmentation) and the growth of tumors along nerves in the skin, brain, and other parts of the body. Most of these tumors, which appear just under the skin, are benign, but people with neurofibromatosis type 1 may develop cancerous tumors that grow along nerves. People with this condition also have an increased risk of developing other malignancies, including cancers of the brain and blood.
>
> The dilemma faced by this young couple was very real, because while Risa's case was mild, there is huge variation in how neurofibromatosis type 1 is expressed. For such a couple, the decision to keep a pregnancy in which the fetus may carry the disease-causing mutation is not so clear.

In light of these issues, I realized that in advising Risa and Edward, it would be important to maintain a nondirective approach. Ultimately, the decision about whether and how to get pregnant, or whether to perform genetic tests during a pregnancy, would be the couple's alone.

"I am very glad you came in to see me today," I began. "Let's start by talking about the nature of your neurofibromatosis and the ways it might present in a child who inherits the mutation that causes this condition. Then, I'd like to learn more about your personal preferences so that I can help clarify your options."

Risa and Edwards's example presents parents who are considering how best to handle the possibility of passing on a specific mutation, one that would have a negative effect on the health of their offspring. But things are not always so cut and dry.

Prenatal molecular testing can reveal a genetic predisposition toward possibly developing a disease that will not emerge until adulthood. This includes age-associated neurological disorders like Parkinson's disease, as well as certain forms of cancer. For instance, prenatal testing may show a mutation in the BRCA1 gene, a marker for breast cancer risk. Should such diseases, which manifest many decades after birth and may never appear at all, be given weight when deciding whether to keep or terminate a pregnancy?

There is no professional consensus on this. However, the authors of this book argue that when prospective parents come to genetic counseling to investigate genetic factors that could affect their children's health, genetic testing for adult-onset conditions should be out of bounds.

In our opinion, a line must be drawn to prevent clinical genetic counseling from devolving into a form of eugenics, which would seek to control human reproduction with the goal of increasing "desirable" heritable characteristics within the population. The noxious corollary of this approach is eugenics' goal of limiting the number of humans with "undesirable" characteristics—an approach that reached its height under the Nazis, who murdered six million Jews, as well as Roma, homosexuals, and others during World War II.

There is another reason to leave adult-onset diseases out of genetic tests conducted during pregnancy. Medical science is advancing every day. Within decades, there may be a cure for the conditions that, at present, we most fear to pass on to our children.

It is impossible to discuss the ethics of genetic testing without entering into the ethical discussions around abortion and the laws that seek to control it. In the United States, the right to an abortion was enshrined in the 1973 Supreme Court decision *Roe v. Wade*, which declared that the Constitution's "right to privacy" included a woman's right to seek out and have an abortion and that this decision was valid throughout the country under federal law. However, in 2022, the Supreme Court overturned *Roe v. Wade*, declaring the decision to be based in a faulty interpretation of the Constitution and relegating decision-making power about abortion policy to the individual states. The rollback of *Roe v. Wade* has rendered abortion illegal or highly restricted in many states, creating inequities in women's access to the procedure that can culminate in endangering women's lives. It also endangers the livelihoods and licensure of those who would prescribe and provide abortions, leading, in the most extreme cases, to prosecution and arrest of the doctors and medical technicians involved.

This ruling has deeply negative ramifications for couples undergoing genetic testing procedures. Freedom of choice has become more limited, as abortion providers in the United States may avoid offering parents their services in order to avoid legal complications. Genetic counselors themselves might refrain from suggesting certain screening procedures, as the path they might suggest in response to a problematic test result—an abortion—might not be available.

As the United States struggles with the ramifications of the 2022 Supreme Court decision, it is instructive to put this in a wider global perspective. A large majority of countries permit abortion under at least some circumstances. Two dozen countries ban abortion entirely. Around 100 countries have some restrictions. However, a complicating factor is the fact that, around the world, legislation concerning the conditions under which an abortion is permitted is often vague, resulting in uncertainty for medical professionals about whether performing certain abortions is legal.

Of course, legal statutes are not the only consideration in making decisions about abortion. While individuals will always decide to what extent they will abide by them, cultural and religious traditions play an important role.

In the Catholic tradition, human life is believed to begin at the moment of conception. Because of this, any deliberate action that destroys a fertilized embryo—such as the disposal of embryos created through IVF or the termination of a pregnancy—is considered morally unacceptable. Judaism has a concept of fetal rights, but they change over time. At very early stages, Jewish law permits the discarding of fertilized eggs that are not implanted, as well as abortion under certain conditions. However, as the pregnancy continues, the fetus is believed to have greater rights, and abortion becomes more restricted.

Islam has three different approaches to defining the beginning of life. The first, similar to Catholicism, posits that life begins when the sperm fertilizes the egg. The second, more amorphous, teaching says that life begins when the soul enters the body. The third tradition considers the implantation of the embryo in the woman's body to be the critical factor. This third approach makes IVF and preimplantation genetic testing a good option for certain, but not all, Muslim parents.

A wider discussion of the legal control of abortions and religious attitudes toward them is beyond the scope of this book. Instead, let's go back to the clinic to describe something that makes it possible, in many cases, for individual at-risk couples to avoid abortion altogether.

> Risa and Edward knew they had a problem. Risa's genetic profile meant that any child they might have would be at risk for neurofibromatosis, which can be life-threatening in many cases.
>
> "The fact that you have come in for a consultation before becoming pregnant puts you in a good position," I said. "You have the option to choose to become pregnant through IVF, in vitro fertilization. This would allow genetic testing to be done on your fertilized eggs so that only a healthy embryo that does not carry the mutated gene is implanted."
>
> I explained the medical and laboratory procedures that ensure that only the healthy early-stage embryos are candidates for implantation into the womb.

Edward seemed stressed and even a bit embarrassed by these technical details. He turned toward his wife, offering a wry smile. "So much for romance," he said.

"It is true that this is not the way young couples envision starting a family," I said. "But it is an excellent option, one that allows you to take preemptive responsibility for your child's health."

Having reached this point in our discussion of the ethics of genetic testing, the reader will not be surprised to learn that preimplantation testing has its issues as well. Because the process aims to choose a healthy early-stage embryo, there will be, by definition, more embryos created than used. The extra embryos may be discarded—an ethical issue in and of itself—or may be frozen and preserved as "orphan" embryos. Are such embryos property, meaning that they can be discarded according to the parents' wishes? Or are they potential human beings with rights of their own?

A dramatic illustration of this ethical minefield comes from Australia. An American couple, Mario and Elsa Rios, flew to Melbourne in 1981 to seek the services of a fertility program at the Queen Victoria Medical Center. Three of Elsa's eggs were extracted and fertilized, but only one was implanted. The pregnancy resulted in a miscarriage. The couple returned to America, but their two other fertilized eggs remained in storage at the Australian clinic, frozen in liquid nitrogen.

Two years later, Mario and Elsa died in a plane crash. Not long afterward, their story made headlines when a London paper reported that the couple had a substantial fortune and had left no will. The only heirs to the Rios's fortune, it seemed, were the pair of potentially viable embryos that the couple left behind in the Australian IVF clinic's freezer. As preposterous as it may seem, the clinic was deluged with petitions by women who asked to be impregnated with the Rios's fertilized eggs, in hopes of inheriting the late couple's fortune.

Ultimately, the Victorian authorities ruled against allowing this disturbing turn of events to proceed, and after litigation in the American courts, the Rios's fortune was distributed equally between Mario's son from his first marriage and Elsa's mother. While this case was resolved, the overall question of what should be done with extra fertilized eggs created for IVF remains a burning issue.

The existence of preimplantation genetic testing (PGT) begs another ethical question. Which couples, under what circumstances, should be allowed to use IVF to fine-tune the child to which they hope to give birth?

And the questions keep coming. When the parents' genetic profile indicates a risk that their child could inherit a disorder, should testing be offered only if the disorder is severe? Should the molecular tests conducted prior to implantation be limited to certain severe disorders, or should they be more comprehensive? Who decides what constitutes "severe"? And what happens in cases when testing reveals that the embryo carries mutations associated with multiple conditions that may play out with differing degrees of severity? How should parents analyze the odds and decide what to do with their embryos?

How about parents who wish to choose the sex of their child? Setting aside for a moment China's famously problematic one-child policy, which led to a vast gender imbalance in that country because of a cultural preference for boys, choosing the sex of a child can sometimes be warranted for reasons of health. In cases where a parent carries the mutation for a syndrome known to be more common in boys, the couple may want to choose to have a girl to lower the risk of having a child who inherits the disease.

Autism is more prevalent in boys than in girls. In treating a family that already has one autistic child, should a genetic counselor recommend that a couple do PGT to select for a female embryo? What about the family who already has three girls and is desperate to carry on the family name by having a son?

But of course, choosing a child's sex is a slippery slope to other scenarios. Using genetic testing, it may eventually become possible for parents to define their own perfect child checklist, consisting of characteristics such as height or eye color. This disturbing possibility is not yet a reality, but given this potential for abuse, it is reasonable that there should be some barriers to PGT procedures. PGT should be limited to investigations related to the unborn child's health and viability, not parental preferences.

As we have seen, the path to a healthy pregnancy is not always clear. Genetic tests performed during the course of IVF have an undeniable advantage over fetal testing in the womb; they protect women and their partners from the psychological pain associated with abortion. Many people find it ethically problematic when fertilized embryos not yet implanted in the womb are allowed to die. However, many view this path as preferable to aborting a fetus.

Ultimately, PGT cannot predict the future and determine which embryo, once implanted, will result in a healthy birth. This is especially true with regard to adult-onset disorders. The path toward parenthood and the process of raising children involve accepting a certain amount of uncertainty and risk. Hopefully, medical science will eventually find treatments and cures for the genetic conditions revealed by testing. Until then, we muddle along and, with clinical guidance, make the best, most informed decisions we can.

Chapter 21
The Future of Medical Genetics

Throughout this book, we have followed the path from genes to diagnosis to deeply personal decisions. We have tried to show in this book how fundamental research done in genetic science labs was essential for establishing today's best practices in genetic counseling. But what about tomorrow? How will future discoveries impact how we understand, diagnose, and treat inherited syndromes? Given the speculative nature of this question, how far ahead should we predict? A year? A decade? A century? More?

As authors, we are in the business of science, not science fiction. Therefore, we have assembled some predictions about what genetics and genetic counseling may look like in the near future, based on current scientific knowledge and technological capabilities.

21.1 Genetic Counseling of the Future

The COVID-19 pandemic of 2020 changed the world dramatically, as lockdowns made face-to-face meetings impossible. Digital communication platforms created to deal with this crisis are still with us and can provide new ways to improve both the accessibility and efficiency of genetic counseling.

Today, genetic counseling is based on in-person consultations with specialists. In the future, an increased reliance on virtual telemedicine may benefit patients and families. Such platforms may enable timelier appointments with the limited amount of geneticists and genetic counselors available. They may also help overcome treatment disparities between city dwellers with access to hospitals and patients living in underserved areas. Putting clinical services online may also create opportunities to increase awareness of genetic counseling and to attract patients who otherwise might not seek advice about genetic conditions or family planning.

E. Banne, B.-Z. Shilo, *Dialogue With Our Genes*,
https://doi.org/10.1007/978-981-95-9571-6_21

This change, however, would come with a cost. Increased use of digital platforms for genetic counseling would reduce the intimacy and focus that is such an important part of the patient-clinician relationship. Physical visits to the clinic force patients to listen and take an active part in the conversation. But when conversations can take place anywhere—on the street, at the office, while driving, or while attending to childcare or household tasks—communication can be less effective. Critical details may not sink in or may be quickly forgotten. This makes virtual communication a less desirable alternative to in-person counseling.

On the other hand, the automated information-gathering made possible by virtual platforms could speed up what is now a very long consultation process. Much of a counselor's time is devoted to collecting data about suspected genetic diagnoses, family history, and general and medical backgrounds of the consultee. The need to integrate the data collected and to determine which genetic tests to run leads to extended waiting periods. If general intake data is collected in the same manner for every patient, it seems reasonable that a computer interface could be created, making it possible for the data to be collected, at least in part, by machine.

Imagine this scenario. Before your first visit to the clinic, you go online to fill in answers to a series of questions defined by a professional genetic counselor. Your answers to the questions are recorded in an electronic medical record system for review by a clinician prior to your first in-person or virtual visit. Electronic medical records are already used in medical centers and hospitals, but future improvements to this technology could make genetic counseling more efficient by generating a list of potential diagnoses or even creating a genetic family tree in digital form, based on patients' responses.

It is also conceivable that someday in the future, we could use the automatically generated family trees to create links between previously entered genetic data related to family members, thereby speeding the counselor's progress toward an accurate diagnosis. Although this hypothetical linkage might seem problematic in view of today's standards for patient privacy, such information-sharing might be deemed acceptable in the future.

Future integration of genetic counseling into digital platforms may break down barriers and hierarchies, allowing counselors and consultees to work together. In this scenario, every patient may eventually have access to their own genetic blueprint, stored in its entirety and accessible online. Together, doctors and patients could explore this resource and use it dynamically, analyzing risks for specific diseases, exploring how genetics influence response to medications, or even tailoring lifestyle choices to a unique genetic profile.

Digital interfaces could also improve clinical care by tracking and promoting patients' technical compliance. For example, an automated chat box could contact patients to ask whether they were able to follow the genetic counselor's instructions or if they completed recommended tests.

Such automated follow-up could provide emotional support. Recent advances in an AI approach called *natural language processing* have demonstrated how machines can carry out human-like AI-based conversations, responding appropriately to emotional indicators. For example, if patients have not followed through on recommended procedures, an AI system could ask what happened and offer needed reassurance. If genetic test results reveal a troubling outcome, AI systems could offer help through AI-driven chats that provide a measure of immediate psychological support while offering links to online and community-based resources. These AI-driven interactions can also serve as a bridge to a session with a human genetic counselor.

We do not suggest that digital automation and AI will eliminate the need for human interaction as patients make their way through the genetic counseling process. But it is reasonable to think that advances in these areas will improve access to genetic counseling, increase genetic knowledge, and make testing and diagnosis more efficient.

21.2 Fundamental Genetic Knowledge Set to Increase

Today we know the disease-causing function of approximately 5000 human genes. But additional discoveries are emerging every day; eventually we will know how all 20,000 human genes affect human development and health. This treasure trove of information will give tomorrow's genetic counselors the opportunity to play a much more significant role in disease prediction, prevention, and treatment.

Future scientific research will help us understand the genome more holistically. It will generate the data needed to better understand the 1% of genes that code for proteins and to understand the subtle differences caused by gene variants. Laboratory advances will clarify how altered gene expression leads to altered function and how proteins encoded by these variants function within the cell and in extracellular networks. Yet another challenge that tomorrow's lab advances may help overcome relates to our currently limited understanding of how each gene interacts with, and influences the activity of, other genes. This is especially important when addressing diseases caused by the interplay of several factors.

Progress in computer science and improved computer modeling of biological processes will help tomorrow's scientists characterize proteins involved in complex networks while clarifying the activity of proteins that play more than one functional role in the body. This will make it possible to understand genetics and genetic disease at a level of detail beyond what can be achieved with today's methods.

Finally, laboratory science will push genetic counseling forward by answering questions that scientists have only begun to ask. For example, it may reveal how parts of intertwined chromosomes interact with each other, an interaction that likely influences gene expression but is not currently understood. Over the coming decades, genetic researchers will certainly produce additional breakthroughs, creating new tools and expanded knowledge that will improve the way that genetic counselors do their work.

21.3 A Future (and Necessary) Reduction in Genetic Testing Cost

Despite dreams of the $1 genome, genetic testing is prohibitively expensive. There has been progress—the cost of a human whole genome sequencing decreased from an estimated $1 million in 2007 to $1000 in 2014 and by 2025, to approximately $600—but testing needs to become even cheaper to ensure accessibility and to end unequal access to genetic data.

This hopeful future scenario for cheaper genetic testing will affect health globally by making it possible to open niche markets. Developing countries would particularly benefit greatly from this, as affordable testing would drive the emergence of personalized clinical care tailored for their specific populations.

21.4 A Partner for Genetic Progress: Artificial Intelligence (AI)

As this book is being written, AI systems are still having trouble integrating complex genetic information to achieve accurate medical diagnosis. But things are improving. It is believed that with a small technological leap, it will become commonplace for physicians to use AI in their medical practice.

Future integration of big data and AI machine learning is expected to revolutionize how medical geneticists interpret genetic information. It will be particularly helpful for multifactorial inherited conditions caused by changes in more than one gene. Such disorders—like type 2 diabetes, high blood pressure, and obesity—clearly have a genetic basis, but the multiple gene variants involved trigger subtle network-based dynamics that defy scientists' analytic efforts. The powerful AI algorithms of tomorrow may make such puzzles easier to solve.

Further integration of AI and big data into genetic science will lead to faster, more accurate diagnosis of genetic conditions as well as better patient outcomes. AI is already being used to identify disease-related biomarkers—particular patterns of gene expression—and to sift through the data contained in population-wide electronic medical records. AI-based investigations have revealed important health-related findings, such as successfully identifying which healthy individuals are likely to develop blood cancer sometime in the future. Population-wide genomic databases may soon make it possible to identify disease risk for a range of genetic conditions, helping counselors to formulate preventative treatment plans.

In the future, AI-driven methods could become part of standard screening procedures for all patients, whether or not they choose to consult with a genetic counselor. Such a practice would undoubtedly lead to the early detection of certain hereditary conditions and rare genetic syndromes.

21.5 Waiting in the Wings: New Diagnostic Technologies

In the coming decades, deciphering the impact of genetic variations will benefit from the creation of new tools for genetic testing. Based on these tools, more and more people will achieve an exact diagnosis of their medical condition, rather than a working diagnosis, a phrase clinicians use to indicate they don't really know what is wrong with a patient but are doing their best to find out.

A significant technology that may have great future impact is the liquid biopsy. The idea for diagnosing cancer using liquid biopsies arose from the fact that blood contains free-floating DNA as well as proteins shed by dead blood cells in healthy people and, in cancer patients, by dead tumor cells as well. Once the price has lowered enough to make this part of regular screening, liquid biopsies may become the go-to method for identifying the genetic variations that can lead to developmental syndromes and genetic diseases.

21.6 Improving Genetic Disease Treatment

Identifying potential drugs, testing them, and getting them approved by the FDA can take decades and billions of dollars. Up-and-coming technologies may usher in new solutions for major diseases, as well as for rare conditions that tend to be ignored by pharmaceutical companies because they prefer to invest in blockbuster drugs rather than medicines needed for small groups of patients.

Current trends in disease therapeutics have the potential to positively impact the way we diagnose and treat genetic conditions in the future.

Powerful computer programs can already solve the 3-D structure of complex proteins and help design synthetic proteins with beneficial functions. More powerful AI-enabled programs will go beyond these achievements, creating platforms in which the entire pathway of interacting proteins can be simulated and studied. Such models could be used to test potential drug compounds to see if they can be used to treat genetic diseases.

Gene therapy is a long-hoped-for method of replacing disease-related genetic variants with healthy sequences that do not lead to disease. In recent years, this approach has achieved a certain level of practical application. Hemophilia A, for example, has been successfully treated with a one-time infusion that delivers a functional gene to replace a defective gene that triggers problems with blood clotting. Gene therapy has also been applied to the treatment of spinal muscular atrophy (SMA), a motor neuron disease associated with progressive muscle weakness and respiratory failure.

In gene therapy, replacement genes are packaged into viruses (vectors) that deliver the genes directly into cells. Today, biological engineers are developing more sophisticated vectors—tiny viral carriers capable of delivering the right gene to the right cells. If successful, this could be a huge leap forward in the treatment of genetic conditions.

Another scientific breakthrough that may impact future clinical practices is *CRISPR-Cas9*. This technology revolutionized the scientific world by making it easy to make small changes to DNA sequences in cells. Using this technology to repair cells in living organisms—first in model systems and then eventually in humans—would be a natural next step for CRISPR. It could potentially be used to treat congenital genetic disorders in utero or at very young age or to cure genetic diseases in adults.

Advances in the study of stem cells—the undifferentiated precursor cells that mature into different kinds of body cells—could also impact the way we treat genetic conditions. A relatively new technology makes it possible to reestablish the stemness of fully differentiated and mature skin cells—to revert fully developed skin cells back to a stem-cell-like state, restoring their ability to become any type of cell in the body. The induced stem cells created by this process could be directed to generate entirely new organs for transplantation procedures, a line of research that could eventually provide treatment options for certain genetic syndromes.

A new kind of vaccine based on messenger RNA (mRNA) was a major medical breakthrough that emerged from the COVID-19 pandemic. It is reasonable to assume that in the future, mRNA technology could be used to target and destroy cells that carry a particular genetic mutation. Already touted as a possible strategy in the fight against cancer, such vaccines may also be put to work fighting genetic disease.

21.7 Future Ethical Challenges

As genetic counseling becomes more integrated into everyday healthcare, it raises profound ethical and societal questions. Will everyone have equal access to the benefits of genetic medical science, or will the field develop in a way that widens existing health disparities? How do we ensure the privacy of such sensitive information in a world where data breaches are all too common?

Despite these challenges, genetic counseling's potential to empower individuals and families is undeniable. It offers a future where people are armed with knowledge about their genetic makeup and can make informed decisions about their health and the health of future generations. As technology and ethics advance side by side, genetic counseling will remain a cornerstone of compassionate, personalized medicine in the years to come.

Glossary

Allele An allele is one of two or more versions of the same gene, found at the same position on a pair of chromosomes—one inherited from each parent.

Amino acid A building block of proteins. It consists of a central carbon atom bonded to a hydrogen atom, an amino group, a carboxyl group, and a unique side chain. There are 20 different amino acids that combine in various ways to form proteins

Amniocentesis A medical procedure where a small amount of amniotic fluid is taken from the sac surrounding a baby during pregnancy. It is used to test for genetic conditions, birth defects, and infections.

Asymptomatic Having a condition or disease but not showing any symptoms or signs of the condition.

Base, genetic One of the four building blocks of DNA: adenine (A), thymine (T), cytosine (C), and guanine (G). These bases pair together in specific combinations (A with T, C with G) to form the structure of DNA.

Base pairs Pairs of nitrogenous bases that connect the two strands of the DNA double helix and give it its structure. In DNA, adenine (A) pairs with thymine (T), and cytosine (C) pairs with guanine (G). These base pairs are held together by hydrogen bonds.

Carrier A person who has one copy of a gene for a recessive genetic condition but does not show symptoms of the condition. A carrier can pass the gene to offspring, who may inherit the condition if they receive another copy of the gene from the other parent.

Carrier testing (screening) A genetic test to determine if a person carries a gene for a recessive genetic disorder, even if they do not show symptoms. Carrier testing helps assess the risk of passing the condition to offspring.

Cell The basic unit of life in all living organisms. It is the smallest structure that can carry out all the processes necessary for life, including metabolism, growth, and reproduction.

E. Banne, B.-Z. Shilo, *Dialogue With Our Genes*,
https://doi.org/10.1007/978-981-95-9571-6

Cellular membrane A thin, flexible barrier that surrounds a cell. It controls what enters and exits the cell and helps maintain the cell's structural integrity.

Chorionic villus sampling (CVS) A prenatal test where a small sample of tissue is taken from the placenta to test for genetic disorders in a developing fetus. It can be done earlier than amniocentesis, typically between the 10th and 13th weeks of pregnancy. The test helps detect conditions like Down syndrome, cystic fibrosis, and other genetic issues. The tissue sample is usually obtained either through the cervix (transcervical) or through the abdomen (transabdominal).

Chromosomal microarray A test that looks for changes in chromosomes, such as missing or extra pieces of DNA, to identify genetic conditions or abnormalities.

Chromosome A threadlike structure made of DNA and proteins that carries genetic information in the form of genes. It is found in the nucleus of cells and becomes visible during cell division.

Clinical geneticist A doctor who specializes in diagnosing and treating genetic disorders. Clinical geneticists help patients and families understand inherited conditions, provide genetic testing, and offer guidance on managing and preventing genetic diseases.

Comparative genomic hybridization (CGH) A laboratory technique used to analyze the copy number variations (CNVs) in the genome. It compares the genetic material of a patient to a reference genome to detect differences, such as deletions, duplications, and amplifications of DNA segments. CGH is particularly useful in identifying chromosomal abnormalities that may be linked to diseases or conditions, including cancer.

Cytoplasm The jellylike substance inside a cell, excluding the nucleus. It surrounds the cell's organelles and is where most cellular processes occur, including protein synthesis and energy production.

DNA (Deoxyribonucleic acid) DNA is a molecule that carries the genetic instructions for the growth, development, functioning, and reproduction of all living things. It is the blueprint of life.

DNA microarray A laboratory tool used to detect the expression of thousands of genes at once. It consists of a small chip with a grid of DNA sequences that can bind to complementary DNA or RNA from a sample, allowing scientists to analyze gene activity or identify genetic variations across multiple genes.

Dominant trait A genetic characteristic that will appear in an individual even if only one copy of the gene for that trait is inherited. It masks the effect of a recessive gene if both are present.

Double helix The shape of DNA, consists of two strands twisted around each other, like a spiral staircase. These strands are made up of nucleotides, which store genetic information.

Dysmorphology The study of physical abnormalities or unusual features in body structure, often used to identify genetic conditions or syndromes.

Egg The female reproductive cell that combines with the male sperm during fertilization to form a new organism. It carries half the genetic material needed for reproduction.

Embryo An early stage of development in a fertilized egg. It begins after fertilization and continues as cells divide and develop into different tissues and organs, eventually becoming a fetus.

Enzyme A protein that speeds up chemical reactions in the body. Enzymes help break down substances, build molecules, and control various biological processes without being consumed in the process.

Eugenics The idea or practice of improving the genetic quality of a human population by selectively breeding people with desirable traits and discouraging those with undesirable traits. It has been widely criticized for its unethical applications.

Evolution The process by which living organisms change over time through variations in their traits. These changes are passed down through generations, and natural selection helps determine which traits are beneficial for survival.

Exome sequencing A genetic test that focuses on sequencing the exons, the parts of genes that code for proteins. It helps identify mutations and genetic variations that may be linked to diseases or health conditions.

Family tree technique A method used to visualize and trace the relationships and genetic traits passed down through multiple generations of a family. Building a family tree helps genetic counselors study how traits or conditions may have been inherited.

Gene A small section of DNA that contains the instructions for making a specific protein. It determines a particular trait or function in an organism.

Gene panel A test that examines multiple genes at once to detect mutations or genetic variations associated with certain diseases or conditions. It helps to identify genetic risk factors in a more targeted way than whole genome sequencing.

Geneticist In medicine, a geneticist is a physician-scientist who studies genes and how hereditary traits are transmitted from one generation to the next. Geneticists seek to understand and diagnose genetic conditions.

Genetics The study of how traits and characteristics are passed from parents to offspring through genes. It explains how organisms inherit physical and behavioral traits.

Genome The complete set of DNA in an organism, including all its genes. The genome contains all the instructions needed for the organism to grow, develop, and function.

Genotype The genetic makeup of an organism, consisting of the genes inherited from both parents. The genotype determines traits and characteristics, like eye color or susceptibility to certain diseases.

Histone A type of protein found in the nucleus of eukaryotic cells (cells that have a nucleus). Histones play a crucial role in organizing DNA into a compact structure. They help regulate gene expression by controlling how tightly or loosely the DNA is packaged, which influences which genes will be expressed.

Homologue A gene, chromosome, or protein that is similar in structure and function across different species or within the same organism. Homologues arise from a common ancestor and share a similar evolutionary origin.

Homozygous Having two identical alleles for a particular gene, one inherited from each parent. For example, an individual can be homozygous for a dominant or recessive trait if both alleles are the same.

Human Genome Project A global scientific effort to map and understand all the genes in human DNA. Completed in 2003, it identified the sequence of about 3 billion DNA base pairs and all the genes in the human genome, helping scientists learn more about health, disease, and evolution.

Hybridization A process in which two complementary strands of nucleic acids (such as DNA or RNA) bind together to form a double-stranded structure. This process is used in various tests, such as in hybridization assays, where labeled nucleic acid probes are used to identify specific sequences in a sample.

In vitro fertilization (IVF) A medical procedure where an egg is fertilized by sperm outside the body in a laboratory. The fertilized egg (embryo) is then implanted into the uterus to achieve pregnancy.

Junk DNA Outdated term. Relates to parts of the DNA that do not appear to have a direct function in coding for proteins. For many years, these regions were thought to be useless, but recent research shows that they may play important roles in regulating genes and maintaining the structure of the genome.

Karyotype A picture or diagram of a person's chromosomes, arranged by size and shape.

Karyotyping The process of analyzing the full set of chromosomes to detect genetic abnormalities, such as extra or missing chromosomes.

Mendel, Gregor Johann An Austrian scientist and monk who is known as the "Father of Genetics." He discovered the basic principles of heredity through his experiments with pea plants, which laid the foundation for modern genetics.

Metabolic tests Medical tests that measure how the body processes substances like sugars, proteins, and fats. These tests help diagnose disorders related to metabolism, such as diabetes and inherited metabolic conditions.

Molecule A group of two or more atoms bonded together. Molecules are the smallest units of chemical compounds and can consist of elements like water (H_2O) or complex substances like proteins.

Mutation A change in the DNA sequence of a gene. Mutations can occur naturally or be caused by environmental factors. They may lead to variations in traits and diseases or may have no noticeable effect at all.

Neuron A type of cell that is the basic building block of the nervous system. It transmits electrical signals throughout the body, allowing for communication between different parts of the brain, spinal cord, and other organs. Neurons are essential for sensory perception, motor coordination, and cognition.

Next-generation sequencing (NGS) An advanced technology used to quickly and accurately determine the sequence of DNA or RNA. It allows scientists to study entire genomes, specific genes, or genetic variations in detail.

Nucleosome The basic unit of DNA packaging in eukaryotic cells (cells that have a nucleus). Nucleosomes consist of a segment of DNA wrapped around a core of histone proteins. Nucleosomes help organize the DNA into a compact structure, which allows it to fit inside the nucleus of the cell and plays a role in regulating gene expression by controlling how tightly or loosely the DNA is packed.

Nucleotide Also known as a base, a nucleotide is a basic building block of DNA and RNA. It consists of three components: a sugar, a phosphate group, and a nitrogenous base (adenine, thymine, cytosine, or guanine in DNA; uracil replaces thymine in RNA). The chemical bonds between nucleotides form the structure of genetic material.

Nucleus The control center of a cell. The nucleus contains the cell's genetic material (DNA) and coordinates activities such as growth, metabolism, protein synthesis, and cell division.

Oncogene A gene that has the potential to cause cancer. When mutated or overexpressed, it can lead to uncontrolled cell growth and division, contributing to cancer onset.

Percutaneous umbilical blood sampling A medical procedure used to collect a sample of blood from the umbilical cord of a fetus. It is typically performed when certain fetal conditions need to be diagnosed or assessed, such as genetic disorders, infections, and blood disorders. The procedure is done by inserting a needle through the mother's abdomen into the umbilical cord, guided by ultrasound.

Phenotype The observable physical traits or characteristics of an organism, such as eye color, height, and blood type, which result from the interaction of its genotype and the environment.

Phenotypic pedigree A diagram that shows the inheritance of observable traits or characteristics across generations in a family. It is used to trace the pattern of traits and can help identify the genetic basis of certain characteristics or conditions.

Polymerase chain reaction (PCR) A laboratory technique used to make millions of copies of a specific segment of DNA. It allows scientists to analyze and study small amounts of DNA in detail.

Preimplantation genetic testing (PGT) A medical procedure that helps identify embryos that are at risk for inherited diseases, chromosomal abnormalities (like Down syndrome), and other genetic disorders. PGT allows for the selection of healthy embryos for implantation, reducing the risk of genetic conditions being passed on to the child.

Progeny The offspring or descendants of an organism, including children, young animals, and plants produced from seeds.

Proteins Large molecules made up of amino acids. Proteins perform a wide variety of functions in the body, such as building tissues, acting as enzymes to speed up chemical reactions, and supporting immune responses.

RASopathy A group of genetic disorders caused by mutations in the RAS/MAPK pathway, which affects cell growth and development. These conditions can lead to developmental delays, heart defects, and distinct physical features.

RAS pathway A signaling pathway inside cells that regulates important functions such as cell growth, division, and survival. The RAS pathway involves a series of proteins that transmit signals from receptors on the cell surface to the cell's nucleus, influencing gene expression.

Recessive trait A genetic characteristic that only appears when an individual inherits two copies of the recessive gene, one from each parent. It is hidden if a dominant gene is present.

Ribosome A cellular structure responsible for protein synthesis. It reads the information in mRNA and links amino acids together to form proteins, the biological molecules that carry out cellular functions.

RNA (Ribonucleic acid) RNA is a molecule like DNA but is single-stranded. It plays a key role in translating genetic information from DNA into proteins and helps in various biological processes such as gene expression and regulation.

Sanger sequencing A method used to determine the exact sequence of nucleotides in a DNA molecule. It is commonly used for smaller-scale genetic analysis, such as sequencing a specific gene or region of DNA.

Sequencing The process of determining the exact order of the bases (also called nucleotides) in a DNA molecule. Sequencing helps scientists understand genetic information, study genes, and identify mutations.

Signal transduction The process by which a cell responds to external signals, such as hormones or nutrients, through a series of molecular events. These events lead to a cellular response, such as changes in gene expression or cellular activity. Signal transduction is how cells communicate and adapt to their environment.

Somatic Cells that are not part of the reproductive system, such as those that make up the body's tissues and organs. Somatic cells have two copies of each chromosome, unlike gametes (egg and sperm cells), which have only one.

Sperm The male reproductive cell responsible for fertilizing the female egg during reproduction. It carries half the genetic material needed to form a new organism.

Synapse The junction or connection between two neurons, or between a neuron and a muscle cell. Synapses are where the electrical signals (action potentials) are transmitted from one neuron to another or to a target cell. The synapse plays a key role in the transmission of nerve impulses across the nervous system.

Trisomy A genetic condition where a person has three copies of a chromosome instead of the usual two. This can lead to developmental and physical differences. An example is Down syndrome, which stems from trisomy of chromosome 21.

Tumor suppressor A gene that helps regulate cell growth and repair DNA; loss of its function can contribute to cancer.

Variant of unknown significance (VUS) A genetic change found during testing whose impact on health or disease is not yet known. Further research is needed to determine if it is harmful or harmless.

Whole Exome Sequencing A genetic test that analyzes all the protein-coding parts of genes (called exons) in a person's DNA. Whole exome sequencing helps identify genetic changes that may cause diseases or disorders.

X-linked Genetic traits or conditions that are associated with genes located on the X chromosome. These traits are often inherited differently in males and females, because males have only one X chromosome, while females have two.

GPSR Compliance

The European Union's (EU) General Product Safety Regulation (GPSR) is a set of rules that requires consumer products to be safe and our obligations to ensure this.

If you have any concerns about our products, you can contact us on ProductSafety@springernature.com

In case Publisher is established outside the EU, the EU authorized representative is:

Springer Nature Customer Service Center GmbH
Europaplatz 3
69115 Heidelberg, Germany

Batch number: 10889560

Printed by Printforce, the Netherlands